Dedication

This writing is dedicated to Norma Jean, my wife of 59 years, for her patience and time alone during the writing. She is my life. I must also thank my son Jon for the overall formatting and production of the finished product. It could not have been done without you, son.

Federal Government Proposal Writing

A Simplified Teaching For Beginners and Small Businesses

TABLE OF CONTENTS

Section	*Title*	*Page*

1.   HOW IT ALL BEGINS 8
    1.1   Marketing 8
        1.1.1   Marketing Plan 8
        1.1.1.1   Working Outside of the Box 9
        1.1.2   Basic Events Leading To Proposal Development.............9
        1.1.3   The Bid/No-Bid Decision Procedure 9
        1.1.3.1   Overview..10
        1.1.3.2   The Desire to Bid and the Discipline Not to Bid 10
    1.2   THE BID/NO-BID DECISION 13

2.   THE BID DECISION IS REACHED...15
    2.1   Readiness to Respond 15

3.   GETTING DOWN TO BUSINESS – THE REQUEST FOR PROPOSALS 16
    3.1   The ABC's of the Request for Proposal 17
    3.2   An Approach to Reading the Document 20

4.   PROPOSAL KICK-OFF...21
    4.1   Let's Get Started 21
    4.2   The Team 21
        4.2.1   Business Development Manager....................................22
        4.2.2   Capture Manager 22
        4.2.3   Proposal Manager 23
        4.2.4   Proposal Coordinator 23
        4.2.5   Volume Leaders...23
        4.2.6   Proposal Writers..24
        4.2.7   Proposal Review Teams...24

5.   THE PROPOSAL DEVELOPMENT PROCESS 25
    5.1   The Proposal Schedule...25
        5.1.1   Pre-RFP--Bid/No Bid Decision 26
        5.1.2   Proposal Kickoff 26
        5.1.3   Proposal Development 26
        5.1.4   The Internal Review & Modification Process 27
        5.1.5   Gold Team Executive Review 28
        5.1.6   Proposal Submittal 28

6.   RFP DISTRIBUTION 29

7.   PROPOSAL OUTLINE AND CROSS-REFERENCE MATRIX 31

8.   STORYBOARDS 34

9.   THEMES..37

10. DISCRIMINATORS AND OTHER ZINGERS...38
    10.1   Discriminators..38
    10.2   AHA!s...38
    10.3   Ghost Stories (Or Just Ghosts)..38

11. THE PROPOSAL – GETTING DOWN TO BUSINESS 40
    11.1   Executive Summary 40

12. TECHNICAL SECTION OR VOLUME 42
    12.1   Writing Guidelines...43
    12.2   Writing Samples...44
    12.3   Win Themes and Discriminators 44
           12.3.1   Example 1 44
           12.3.2   Example 2 44

13. SAMPLE PROPOSAL SECTIONS 45
    13.1   Past Performance 45
    13.2   TRANSITION PLAN - Fictitious Sample Proposal Section............47
    13.3   On-Going Recruiting and Staffing..52

14. A MANAGEMENT SECTION OR VOLUME 56
    14.1   Program and Task Order Management 57
           14.1.1   Program Management 57

15. TASK ORDER PLANNING 58

16. RESUMES 64
    16.1   Personalize the Resumes...64

17. REVIEWS..66
    17.1   The Yellow Team 66
    17.2   The Blue Team...66
    17.3   The Pink Team...66
    17.4   The Red Team..67
    17.5   The Gold Team Review 69

18. THE COST PROPOSAL 70
    18.1   Element Structure...70
           18.1.1   Direct Labor...70
           18.1.2   Overhead..70
           18.1.3   Other Direct Costs..70
           18.1.4   General & Administrative 70
           18.1.5   Fee...70
    18.2   Types of Government Contracts 71
           18.2.1   Fixed Price Contracts..71
           18.2.2   Cost-Plus Contracts..72
           18.2.3   Labor Hour Contracts 72
    18.3   The Cost Proposal 72

19.   CONTRACT SET-ASIDES 74
    19.1   Small Business Act 74
        19.1.1   The 8(a) Business Development Program 74
        19.1.2   Small Disadvantaged Business 75
        19.1.3   Women-Owned Small Business 75
        19.1.4   Service-Disabled Veteran-Owned Businesses...................76
        19.1.5   HUBZone Procurements 77
    19.2   Determining Business Size 77
    19.3   Mandatory Registrations..77
        19.3.1   DUNS Number...77
        19.3.2   The System for Award Management...................................78
        19.3.3   Tax I.D. 78
        19.3.4   Cage Code..78
        19.3.5   Contractor Performance Assessment Reporting System (CPARS).....79
    19.4   Helpful Government Contracting Sites...79

20.   THE COVER LETTER 82

21.   COVER ART BINDERS...83

22.   PROPRIETARY DATA 85

23.   ORAL TECHNICAL PRESENTATIONS 86

24.   POST AWARD DEBRIEFING...88

25.   IN CLOSING...88

APPENDIX A..89

LIST OF FIGURES

Figure	Title	Page
Figure 1 – Overview of the Proposal Philosophy		9
Figure 2 – Proposal Team		22
Figure 3 – The Proposal Development Process		26
Figure 4 – Making the Correct Bid/No-Bid Decision		30
Figure 5 – The Table of Contents Also Includes a Compliance Matrix		32
Figure 6 – Typical Storyboard		35
Figure 7 – A Good Theme Goes to the Heart of the Customer's Problems.		37
Figure 8 – Make it easy on the Reader. Tell him why he should give us the job.		41
Figure 9 – Rules for Writing Proposals That Win.		43
Figure 10 – Task Order Development Process		60
Figure 11 – Task Order Control Resources		61
Figure 12 – A Scoring Process		69
Figure 13 – Cost Proposal		70
Figure 14 – Ensure Pricing is adequate on Fixed Price Contracts		73

LIST OF TABLES

Table	Title	Page
Table 1 – Typical Proposal Reviews		27
Table 2 – Phase-In Transition Team		48
Table 3 – Estimate of Required Phase-In Hours		49
Table 4 – Major Phase-In Actions		49
Table 5 – In-Process Reviews		66

FORWARD

Greetings, and welcome to the world of Federal Government proposal writing. This book is written for individuals who wish to, or who must learn proposal development to win federal contracts. While I recommend that government proposals be written predominantly in the third person style, the reader will find that this document does not follow that rule. I am speaking to you, the reader, on a person to person basis. Since this writing will not be reviewed by an English professor, I will not follow strict third person writing here. Simplicity is my rule in this writing. That being said, let's move on.

For the Writers

OK! So you have been selected to support a proposal development group, hereinafter referred to as the "proposal team". What is this all about? The answer is simple. As a member of your company's staff, you are being asked to help win new business. After all, new business is what it takes for your firm to stay afloat and TO GROW. It makes sense, you must admit. You are being asked to help your firm grow. Growth supports the company and it supports the employee's career growth.

Why Me?

Proposals are complex documents. The proposal development process requires the support of personnel who possess specific knowledge and skills in a variety of disciplines. You have been selected based on your proven skills as acknowledged by your firm. Whether you are a technical staff member or any other member of the company, you may be called upon.

Being selected to a proposal development support effort should be flattering to you. It means that your firm is willing to place its confidence in you to help win new business. It means that you are important to the company. Think in terms of career growth.

While proposal writing is a complex operation, it can be and is broken down into relatively simple writing tasks. These are later integrated with other tasks to form the complex document used to acquire new business for your firm. This document explains the proposal development process.

Proposal development is a team operation. Each individual becomes a proposal team member. The proposal team is simply a "we're all in this together" syndrome. The proposal team is comprised of members from multiple areas of the firm, each with varying skills. Many established firms employ a specific staff of proposal writers. Their job is to contribute writing skills with specific knowledge of the overall business of the firm. It is important to understand that these are largely not "technical" personnel. They are proposal writers. The technical aspects of the proposal response are frequently written by technical personnel, outside of their normal tasks of working within a client contract. Typically, it is the proposal team's job to put as little pressure on these technical experts as possible. The technical personnel provide the technical substance of the proposal topic which is then used by the proposal writing staff to move the writing to final "proposaleeze" language. Some firms hire temporary consultants to provide the technical substance.

The result of this team process is a finished technical, management and cost document suitable for delivering to the Government client, with the intent of winning a new contract. This is the way Government contractors grow their businesses.

The document you are now reading has been written to make all readers become qualified proposal writers.

What you will learn

This writing covers all aspects of Federal Government technical and management proposal writing, with emphasis on proposal elements typically assigned to the technical staff. The price proposal is also presented in terms that can be understood by the beginner. Relatively few technical writers are members of the costing team. This process is typically set aside for the firm's financial management and upper management personnel. As a writer, you will participate in the development of written segments that when combined with writings of other team members will result in your firm's technical, management and price bid for a contract. Your team will produce a written book which may be printed or may be electronically transmitted to your prospective client, the Federal Government.

Through this material you will learn the overall proposal development process. While we are concentrating on individual writing tasks, it is most important that you, the student, understand how your work integrates with the overall process and how it all comes together. This will provide you with very valuable knowledge of the overall process, information that you may use today and for tomorrows to come.

You will also see an overview of the entire business acquisition process, not only the proposal writing segment. You will learn how business acquisition segments blend to form an integrated successful approach to staying afloat in the complex world of Government contracting. Through reading this material, you will learn:

- ➢ Marketing Prelude to Proposal Writing
- ➢ U.S. Government Request for Proposal (RFP) Contents
- ➢ The Proposal Team Structure Roles and Responsibilities
- ➢ Explanation of Important Proposal Sections for writers
- ➢ Table of Contents Integration with various RFP sections
- ➢ Proposal Writing
- ➢ Specific Proposal Section sample writings
- ➢ Proposal Review Processes
- ➢ Post Submittal Elements

This writing is also for the corporate management team. As a corporate owner or business management individual, you must be aware that the writing of proposals not only includes the technical staff, but the management staff as well. It is important that you work with your technical staff leadership to establish the team that will develop the final document. The management staff required to prepare, or direct the preparation of appropriate sections of the proposal typically include members of the finance department, legal department and corporate officers who can commit the firm to contract performance and to satisfy all contract requirements. Proposals are a corporate team effort.

What this document is

This document is directed to the new (or nearly new) proposal writing candidates. It is a teaching document, not a reference manual. The assumption is that the reader is inexperienced in proposal writing to win government contracts. The intent of this writing is to provide readers with a broad understanding of proposal development, while focusing on the elements he or she is likely to encounter as a draftee from the technical or management staff. The reader will come away from this learning effort with a solid understanding of the proposal development process and will be a potential member of the team that will prepare government proposals. The primary focus of this reading is an understanding of tasks that are most likely to be assigned to members of the technical staff. All sections of the Government solicitation are identified and discussed. Because of this breadth, non-technical or administrative trainees may also benefit from the reading. The reader will understand the structure of Government proposals, the basics of which rarely change. This material will be specifically helpful to individuals who have limited proposal writing experience in the federal Government arena.

What this document is not

This is not a document that proposes a highly structured process that is for sale by its developer. It is intended to educate the reader on the elements comprising the government's Request for Proposal, and how the selected writing team can complete their job in developing a responsive proposal through the application of proven processes that sell the company's ability to meet the Government's stated needs.

The key to this document is simplicity, not deep detail. It is intended to prepare individuals with the knowledge needed to support a proposal team. The reader will not be lost in technical detail.

In no sense is this writing intended to replace or correct Government RFPs. The reader must follow the specific instructions of the RFP in producing his or her work. The Government RFP instructions must be followed to the letter. This is a teaching document. Any conflicts between this teaching and the government proposal instructions must always fall in favor of the government's instructions.

1. HOW IT ALL BEGINS

Well before any thought is given to proposal writing, businesses begin at the corporate base, or vision. What business does the corporation wish to be in? Typically, businesses begin with owners who have a vision and experience within the field/s related to the vision, although this is not always the case. In the beginning, these owners do much of their own marketing due largely to their inability to finance marketing personnel. Eventually the business grows. There are multiple paths that may be taken to achieve growth. Many times growth occurs with a subcontract from a more senior firm with related experience. The company grows from there. More staff is hired, contracts are renewed, and new contracts are awarded directly to the firm.

Hopefully, as the firm becomes established, it develops a corporate Business Development Plan. This plan identifies the products and/or services that the firm will provide, and identifies potential clients, in this case within the Federal Government. Enter the marketing function. It all begins here.

1.1 Marketing

What do we mean when we talk about marketing? Simply put, we are referring to "the selling of our products, capabilities and experience to meet the customer's needs". How is this done? Basically, it requires that your firm be introduced very early on to potential clients who will be in need of the products and/or services offered by your firm. This is a difficult job and needs to done early in the business cycle because once the client introduces a Request for Proposal (RFP), they are no longer able to speak to interested contractors about their needs. The RFP says it all.

Marketing begins very early in the acquisition life cycle of the client's need. Contractors must first survey government clients and their projected needs. An important aspect of this endeavor is for the contractor to become aware of clients who may need the products and/or services that the firm provides. Businesses conferences are critical to gaining this knowledge and achieving an initial recognition by many potential clients of your firm's capabilities. This is a relatively simple, yet time consuming and sometimes frustrating concept, but it must be performed in a repetitive manner at many conferences.

Except for small procurements, the upcoming needs of a potential client are documented within a Federal department's acquisition plan. The marketing element of any federal contractor must acquire future procurement information from agencies that it wishes to work with. The results of this research are utilized in the processes that follow.

1.1.1 Marketing Plan

Nothing is more important than developing and executing a comprehensive marketing plan. This is an ever changing plan used to gather the data, information, and intelligence on upcoming client needs. The plan is typically developed by the marketing staff with technical assistance from the staff. The plan is a projection of the firm's intentions to acquire contracting agreements from government clients. All contacts with potential clients appearing in the marketing plan are documented within the plan. It is a living research tool used to guide the firm's business access efforts.

Specific details and processes of learning which potential client needs what products or services is a complex marketing function and is covered within this document. The message for readers is that proposals submitted to Government agencies without the up-forward marketing activities having been performed, have little chance of winning.

1.1.1.1 Working Outside of the Box

One fact must be observed. No small business marketing staff can possibly track all upcoming procurements. There will be many times when an RFP will be issued from a government agency for which your firm feels qualified to perform. The problem here is that your firm is not known to the potential client. That makes it a time of making a bid/no-bid decision between your corporate management and marketing personnel. This occurs frequently. As a writer, you should be aware that this may occur. When it does, and the firm decides to bid, you the writer will have a more difficult job. Enough said.

1.1.2 Basic Events Leading To Proposal Development

- Development of the corporate Business Development Plan
  - Identifies projected opportunities
  - Identifies potential clients
- Marketing Staff cultivates these and other opportunities
  - Visits to client sites
  - Distribution of Corporate Marketing Material
  - Introduction of corporate Senior Technical Staff to potential clients
  - Attending and participating at Business Networking Conferences
- Tracking the Government's acquisition process for expected solicitations
- Monitoring the specific elements of change within released solicitations
- Making continuation decisions based on unexpected RFP element changes

---

Overview of the Proposal Philosophy

- **Proposals are a key element in the overall Marketing Process**
- **They are the primary vehicle the prospective client uses to evaluate our company and to award us with new work.**
- **The Proposals you develop must be:**
  - *Thorough* - **They must respond faithfully to the client's requirements**
  - *Honest* - **We can embellish but we cannot lie**
  - *Convincing* - **They are a SALES DOCUMENT**

Figure 1 – Overview of the Proposal Philosophy

1.1.3 The Bid/No-Bid Decision Procedure

The Bid/No-Bid process is an analytical approach to selecting and applying opportunity-specific criteria for making Bid/No-Bid decisions. Although the actual Bid/No-Bid process is a quantitative one, the events leading to the decision must include qualitative assessments and information gathering. This procedure addresses the on-going evaluation and qualification of the opportunity during the marketing phase.

While the intent of a marketing effort is to gather the information needed to write a winning proposal and to "position" ourselves using the customer's eyes as the contractor he wants, proposal development efforts should not be authorized until a bid decision is made. While there are many reasons to bid, and even more reasons not to bid, the fact of the matter is that limited proposal resources are available to contractors at any given time. Thus, we must continually assess and reassess our marketing efforts so as to favor those procurements where the risk/reward ratio is in our favor. We must choose our battles carefully, and to save our ammunition (B&P funds and personnel resources) for those procurements

1.1.3.1   Overview

This procedure is designed for flexibility. Thus, a step-by-step process is not imposed on the marketing team in its bid/no-bid recommendations. Rather, a series of tests concerning the cultivation of opportunities versus sound bid decision criteria is presented. These tests in the form of pointed questions concerning the firm's strategic and tactical position relative to a given opportunity culminate in an evaluation process designed to aid in the final decision. During the marketing stage, well before the release of an RFP, our mentality will be that the answers to these questions are the single-most vital determination of our eventual bid posture.

While it should be your firm's intent to bid only those opportunities that have been marketed, the procedure recognizes that on occasion a bid decision will be made even though the answers to the questions are not favorable. These are largely executive decisions.

1.1.3.2   The Desire to Bid and the Discipline Not to Bid

Proposals are the lifeblood of any organization that relies on marketing to the Federal Government for its success. Thus, there is a standing desire to bid on as many contracts as possible. However, since the number of proposals is directly proportionate to the available financial and writing resources, we must choose our opportunities carefully and make bid decisions for the right reasons.

The Process

Your marketing team should continually identify potential business opportunities. Each opportunity is synopsized and discussed at periodic marketing meetings. These will either be qualified or disqualified as new information is learned. Those that are disqualified are discarded. Opportunities that are qualified remain in the monitoring pipeline for future potential interest. During the course of the development phase of opportunities, many events may trigger their disqualification. Only the relatively small percentage of opportunities reaches the Bid/No-Bid stages.

Gathering Intelligence

A substantial element of the marketing process is the research that must be performed. Your firm needs information on contracts, client organization, and information on what the Government does now to acquire the products and/or services that it will be purchasing through an upcoming procurement. The marketing staff has a vital job to perform. Before ever getting to the point where you, the potential proposal writer must write to sell, the marketing staff must convince the corporate leaders that the opportunity is both a good one for the firm, and that the job is winnable. This can be a considerable undertaking when one realizes that the marketing personnel are looking at multiple potential opportunities. Intelligence must be gathered in a manner that hopefully does not divulge to the competition that your firm is interested in the opportunity. That is not a simple task. Remember, that you a potential writer must have as much information as possible regarding the client's needs. This information will allow you to write in a more informative manner come proposal time. As a note, remember that your firm has a marketing and proposal writing budget. Funds and the efforts of the proposal writing personnel must not be wasted on opportunities that your firm cannot win.

Information Regarding the Client

Just who and where does this potential client fit into the Government organization, is vital information. Your firm needs to know why this procurement exists. This is a vital consideration. The "what" of the procurement is also vital. This information leads to the answer to: "is our firm qualified?" If the answer is "no", interest must be dropped. This process is a continual stream of decisions to be made along the marketing cycle leading to a bid/no-bid decision. A no-bid decision may be reached through a "no" answer to any of the following:

> ➤ Is our firm qualified to perform?
> ➤ Are our qualifications convincing?
> ➤ Have we become known to the prospective client?
> ➤ Is the client satisfied with the incumbent contractor?
> ➤ Will substantial sub-contracting be required of our firm but not the incumbent?

We must convince the prospective client of our capability to perform well in the up-coming procurement. This means that "we" must visit the potential client on multiple occasions to learn about client needs. Throughout this process we must convince the client of our qualifications. The client must be knowledgeable of our firm and its ability to satisfy their needs. Later visits to the client should include key members of the technical staff so that the client becomes aware of your firm's qualifications and realizes that your firm can perform well to satisfy the Government's requirements.

Bidding Against an Incumbent Contractor

If there is an incumbent contractor, the task of winning may be more difficult. Much of this situation depends on the client's satisfaction with the incumbent. Making this determination is most important. If the client is satisfied with the incumbent, it may be difficult to win based on your technical proposal. Cost may also be considered, so don't close the door on the bid quite yet. There may also be other ways that your firm may produce a winning proposal. Through this approach we convince the client that they may have the best of both worlds.

Commitments from Incumbent Personnel

It is at this point that your marketing team enters the world of incumbent personnel recruiting for contracts that are labor oriented. This may be achieved through a careful approach to remaining ethical while getting the job done. It should not be surprising that incumbent contractor personnel realize that their firm's contract with the client will be coming up for rebid. While they may be reminded by their firm that they should not talk to potential bidders concerning the re-bid, experience has overwhelmingly shown that the incumbent personnel must look out for themselves. Enter the condition of career assurance. Incumbent contractor personnel regularly apply to multiple contract bidders to achieve some assurance that they will not lose their jobs should the incumbent fail to win.

This situation leads to an excellent opportunity for your firm to produce commitments from incumbent technical person to your firm in the event that your proposal is the winner;

Care to Be Taken

Should your firm decide to recruit incumbent personnel, it is important to attract the correct individuals. Key Personnel are the key words. For instance, ask the question "Why would we offer a win-based position to an individual who has not produced that well in the eyes of the client?" Your team must perform research, and more research. That's marketing's job.

There are multiple ways in which incumbent personnel may be recruited for inclusion in your proposal. Foremost is the support of a technical recruiter or "Head Hunter", as they are known. For a fee based on an actual hire, these individuals may help in a significant manner. As a cost conscious firm you may wish to attract only a single key member of the incumbent's staff in this manner – can you say Project Manager? Acquiring an "if-win" employment commitment from this individual or someone else of a reasonably high position in the contract is a key to success. By acquiring a recruiter on a contingency basis, you are free to talk with the individual that is provided.

The benefits of communicating with a key member of the incumbent staff are apparent. Beyond the obvious benefit of presenting this individual's resume in your proposal, he or she can help in your recruitment phase-in effort should your firm win the contract. In that situation, the individual breaks no rules prior to award related to contractor "inside information". This is an ethical approach to staffing your start-up personnel. Your key individual will provide valuable assistance upon award in recruiting only those individuals from the existing staff who perform well. This approach sets forth a win/win situation for your firm and your new Government client.

Alternate Approach

An alternate approach to acquiring commitments from the incumbent staff during the proposal writing phase is to place a help wanted advertisement in a local newspaper. This approach is a bit less desirable in that it adds your firm to the list of "known bidders". Even so, many firms frequently use this approach.

Additional Alternate Approach

An additional approach is to conduct a recruiting "seminar". This action results from the same newspaper advertisement as mentioned above. What sets this approach apart from others is that all potential employees who respond are invited to an after-hours seminar where they are given a presentation on your firm's history and capability. This is a best foot forward approach to acquiring resumes and the individual's written permission to use his or her resume in your proposal on an if-win hiring basis. It takes work to win contracts.

1.2   THE BID/NO-BID DECISION

And so the prospective client's RFP has been issued. The importance of the research and preparation as previously defined now comes into play. In addition to the material contained in the RFP, the additional marketing information that the firm has acquired is vital in the final bid/no-bid review and decision. This information should largely have been acquired before the RFP is issued, but there is always the possibility that additional data will come to light at any time. We have answered the following in-house questions with a reasonably comfortable "Yes":

- Are we capable of performing?

- Does our firm have any credible advantages?

- Have we presented our capabilities to this client?

- Has the client recognized our capabilities?

- Do we believe that we may defeat an incumbent contractor?

The Final Bid/No-Bid Review Meeting Is The Essence Of The Bid/No-Bid Decision. It Conveys The Following Factors:

- The desirability of obtaining the contract;

- The probability of winning and overcoming any identified risks inherent in our bid strategy;

- The ability to perform the work, if we win, in a credible fashion and in a manner that reflects well on the firm.

Additional Factors Exist. They May Include The Following Questions

- Is there a favored contractor/team? Who?

- Why are they preferred? High performance rating? Total capability? Low Cost? More capable personnel? What makes us think we can unseat them?

- What are the competitions strengths? Technical? Pricing History?

- What are the competition's weaknesses?

- What are the benefits to the company in winning this contract?

- Does this work fit into our strategic plan? (For example, will it provide the technical background and credibility required to bid on future programs?)

- Do we have an "inside salesman" in the customer office? Who is it?

- How much will it cost to bid?

- Do we know the current organization of the customer's office?

- How are we perceived in this office? In the agency? In the service?

- What is our general reputation in the customer's office?

- What are our perceived strengths?

- What are our perceived weaknesses?

Based on the summation of the marketing effort, a Bid/No-Bid decision is made by the corporation. If a No-Bid decision is made, the effort ceases.

Even after a Bid decision is made, the Proposal Manager and Marketing Manager should continue to watch for "gotchas" or new intelligence that may cause your firm to rethink its decision to bid. During the RFP's life cycle, the proposal manager and others close to the effort should constantly re-evaluate the firm's commitment to the proposal. In this way, we can be reasonably sure that we are not throwing good money after bad.

At all stages of the proposal effort, those working on the proposal must be aware of any indicators that suggest that we cannot write a credible (winning) document. These situations may arise through any new or previously unrecognized factors that inhibit our capability claims. These elements may occur as result of new marketing intelligence that may diminish factors that we once believed to be in our firm's favor. These types of factors may result in a no-bid decision, even if we have already begun the writing phase.

2. THE BID DECISION IS REACHED

After a lengthy evaluation of many factors, the firm has arrived at a Bid decision. As can be seen, a lot of research has been performed to reach this point. As earlier stated, this research includes internal determination of the company's capability and credibility to meet the contract requirements. Added to this is hopefully the would-be client's recognition and respect of the firm's abilities gained through an intelligent marketing approach and technical understanding of their needs. The result of these factors hopefully means that the client expects to see your proposal in response to the solicitation. This is hopefully what the firm has achieved. Good work.

2.1 Readiness to Respond

The corporate marketing staff has been busy. It has led the way to prepare the firm to submit a responsive proposal. But they have not been alone. The marketing segment has been supported by multiple elements of the company in this preparation. These include legal review functions to ensure that there are no "gotchas" in the RFP as it relates to the firm. It includes the technical staff leaders to assist in determining how the company's technical skills relate to the client's needs. Technical staff leaders also assist in the determination of who within the technical staff is available, capable and yes, willing to support the development of the proposal. The recruiting element of the firm may also be involved to identify candidates who may be hired to augment the key personnel being proposed. In summary, the proposal process is not a one-person job. We'll learn more about this in the chapter addressing the proposal team.

3. GETTING DOWN TO BUSINESS – THE REQUEST FOR PROPOSALS

Reading a Federal Government RFP...

You don't have to read the whole thing!

> **Do not feel intimidated. It looks complicated for sure. But keep the faith. The following paragraphs will simplify it for you.**

You, the writer, may feel intimidated when you look at a printed copy of an RFP that may be over an inch thick. When you realize how much of its' content is writer-oriented content vs. how much is "officialdom", it's not nearly as depressing. Incidentally, this "officialdom" that I refer to is important information, just not for proposal writers.

Nearly all Federal RFPs look alike in terms of section and chapter content. There are some exceptions to this as related to certain Government clients. Even so, the RFP sections that you are about to see must still be there. The basic format for most Federal RFPs is dictated by the Federal Acquisition Regulation (FAR). The FAR mandates that RFPs be divided into sections A through M. Each RFP section is identified below, with specific emphasis on those of importance to the proposal writing team. To answer the question concerning differences in some RFP formats, simply note the following:

There are 4 major levels of regulations for preparers of RFPs:

➢ Nearly all Federal agencies follow the uniform contract format defined and required by the Federal Acquisition Regulations (FAR) to develop a "Request For Proposal" (RFP)

➢ Agency Supplements to the FAR (DoD, NASA, GSA)

➢ Service Regulations (AFR 70-15, AFR 70-30)

➢ Command Regulations (NAVSEA Source Selection Guide, AFSC 550-23)

Agency, Service, and Command regulations complement and amplify the FAR but do not contradict it.

3.1   The ABC's of the Request for Proposal

➢ *Cover Letter (not always included)*

➢ *Table of Contents (not always included)*

➢ *Section A-Solicitation/Contract Format*

Section A provides very high-level contract identification information. This can be read and understood quickly. It specifies the title of the contract, the procurement number, contract type, period of performance, and the name of an individual with the Government who is the point of contact (POC).

➢ *Section B- Supplies or Services and Prices/Costs*

This section is important to the entire organization. It defines the Contract Line Items (CLINs) and also Subcontract Line Items (SLINS), if any that identify items being purchased (billable items), identifies the period of performance (POP), and identifies the option periods, if in fact this will be a multi-year contract. This section contains the costs that your firm proposes to charge for the products and/or services being proposed. These are the items being purchased by the Government and defines a format for you the contractor to submit your prices. For multi-year contracts, you must submit your projected prices for all following years.

➢ *Section C - Description/Specifications/Work Statement*

This section is of the utmost importance to all involved in performing writing assignments to produce the proposal. It describes (hopefully in substantial detail) the products to be supplied and/or the work efforts to be performed as a result of a contract award. This is the single most important section of the solicitation that is of vital interest to the writing team. It is within this section that you will describe the processes and/or materials that your firm proposes to deliver. Section C will always contain the description of the work to be performed and/or the products to be developed/delivered, or it will refer the reader to an appendix to the RFP where this information may be found. This section may also be associated with other appendices in Section J with other details to enable the proposal teams to become more familiar with the nature and scope of the tasks requested in Section C.

➢ *Section D - Packages and Marking*

Defines how all contract deliverables such as reports and material will be packaged and shipped. This information is important as these instructions may effect costs and raise logistics issues.

➢ *Section E. Inspection and Acceptance*

Describes the process by which the Government will officially accept deliverables and what to do if the work is not accepted. This can also affect costs and identifies tasks you must be prepared to undertake.

➢ *Section F. Deliveries*

Defines how the Government Contracting Officer will control the work performed and how your firm will deliver identified contract items.

➤ *Section G. Contract Administrative Data*

Describes how the Government Contracting Officer and your firm will interact and how information will be exchanged in administration of the contract to ensure both performance and prompt payment.

➤ *Section H. Special Contract Requirements*

Contains a range of special contract requirements important to this particular procurement, such as procedures for managing changes to the original terms of the contract, government furnished equipment (GFE) requirements, and government furnished property (GFP) requirements.

➤ *Section I. Contract Clauses/General Provisions*

This section identifies the Government contract clauses incorporated by reference in the RFP. Each incorporated clause is identified along with its version number and effective date. These clauses will be incorporated into the contract and the bidder agrees to comply with each. This document should certainly be reviewed by the corporate legal staff. These clauses may vary between contracts depending on the Government's needs and will be binding. Proposal development typically does not begin until the legal team reviews them and provides their acceptance. This section requires no writer support.

➤ *Section J. Attachments, Exhibits*

This section identifies the appendices to the RFP. These attachments typically consist of a range of subjects that further amplify overall requirements such as additional technical specifications, floor plans, geographic locations, Government Furnished Equipment (GFE) and others. This is amplifying information that helps the bidders provide a more accurate response. These attachments are referenced here and must be reviewed and analyzed by the proposal development team because it can easily be viewed under some circumstances as an augmentation of the SOW.

➤ *Section K. Representations/Certifications and Statements of Offerors*

This section identifies a range of compliance items concerning your firm. Many of these items are self-certifications using yes/no check boxes. These items cover a wide range of overall corporate compliance including Taxpayer identifications, whether or not you are a minority business, type of corporation submitting the proposal and additional items that confirm the suitability of the bidding firm to compete for the contract at hand. This section requires no proposal writer response. It is typically completed by the company's legal/administrative team.

➤ *Section L. Proposal Preparation Instructions and Other*

This section provides instructions for preparing the proposal. The proposal developer pays strict attention to this section in the outline and preparation of the proposal. These instructions must be followed! Be careful! You will find that these instructions may be augmented through required response information that does not appear in the SOW but appears elsewhere in the RFP. Typically, this describes proposal formatting at a minimum level and outlining. Font size is usually identified along with overall organization. Organization is a key element to this section. Instructions may also contain items such a delivery instructions.

➤ *Section M. Evaluation Criteria*

Section M is vital. It actually identifies the manner in which the Government review team will evaluate the proposal in terms of factors, sub-factors and elements to be used to grade the proposal. These should always link directly to the SOW. It may well be found that sections within the RFP request information that is not specifically identified for review in Section M. Do not allow your team to minimize or ignore factors appearing elsewhere in the RFP but not specifically in Section M. Your team must respond to all sections of the RFP, not simply Section M. If you do not, this would be a catastrophic error.

Once the technical reviews are completed, other members of the Government's evaluation team will combine prices from your cost proposal to form an overall evaluation to determine who wins the award. It is important to understand that the role that price plays in evaluations will vary by the specific acquisition. Cost always counts.

3.2   An Approach to Reading the Document

Typically, as a proposal writer in support of your firm, your need to read the RFP is limited. You have read the meaning of each of the standard RFP sections. The primary element of the RFP that is of typical use to the writer is Section L. Here, the Government has identified how the proposal must (not should) be organized. Here you will also learn elements that the Government lays forth in terms of page count, page layout (margins, fonts, page sizes), media (disk, CD-ROM, video), submission method, and outline/content.

Having learned the organization of the document, you will now want to know how the proposal will be evaluated. Section M shows this in terms of score weighting, the process used to evaluate your proposal, your past performance and "best value" considerations. In learning the primary evaluation elements, you must also be aware that other elements are equally as important in terms of completeness. Failure to address all applicable RFP elements within Section L (inclusive with other sections), will result in your proposal being considered as non-responsive by the Government evaluator.

Next, return to Section C to determine the writing requirements (what your firm must perform and/or supply). When reading Section C, pay particular attention to the requirements set forth. Your goal here is to ensure that the requirements are clear and do not contain contradictions (between requirements as well as Section L and M), and that the requirements are feasible.

You should remain aware that the RFP is prepared and written by different Government employees. There are times when they add "boilerplate" without determining that it should be. This could modify the context of the solicitation. If this occurs, and is not detected by the Contracting Officer, your proposal management staff will more than likely pick this up. You also should learn to detect these situations. When these types of occurrences are detected, the Contracting Officer should be notified for clarification and/or correction.

Finally, understand that your job is to provide a solution to the Government client's stated needs, ***not what you believe is better or best***. The client specifies and you respond.

4.   PROPOSAL KICK-OFF

4.1   Let's Get Started

Every proposal development project should begin with a "Kick-Off Meeting". For small businesses, this may consist of only a few individuals. Small businesses frequently operate in a multi-task mode where individuals perform multiple functions in the proposal development process. This approach is frequently necessitated by the overall un-availability of writers. Of course, the smaller the proposal requirements, the fewer staff members are required even in the case of larger businesses. The lack of writer availability is frequently solved through the acquisition of technical specialists on a consulting agreement. There may also be subcontractors who will be proposed to the client as in the case of a team bid. In those cases, the subcontracting firms may well augment the proposal writing staff, lessening the overall writing burden.

As with any function, the larger the staff, the more difficult the management function becomes. This is particularly true in functions such as proposal development. Why? Because the proposal development function consists of individuals from various segments of the company, each with a distinct role to fulfill in the development of the document which we all wish to be a work of art. The need is to get each member of the proposal development team to perform their individual functions in a consistent, timely and professional manner.

4.2   The Team

Figure 2 identifies a logical proposal development team. Please note these are organized by functions to be performed. A single individual may in the case of small companies, (or small proposals) perform more than one function.

The key word is teamwork. The proposal is to be developed as a single minded document. We achieve this through a number of factors. Foremost among these are "themes" and "sub themes". We'll look at these and other team elements later. For now, let's talk about the typical team members and their areas of focus. We start with the Capture Manager and the Business Development Manager. Because of the need to promote marketing as separate from actual proposal development, these positions will be discussed first. Note also, that these positions are frequently combined and assigned to a specific individual, especially within small businesses.

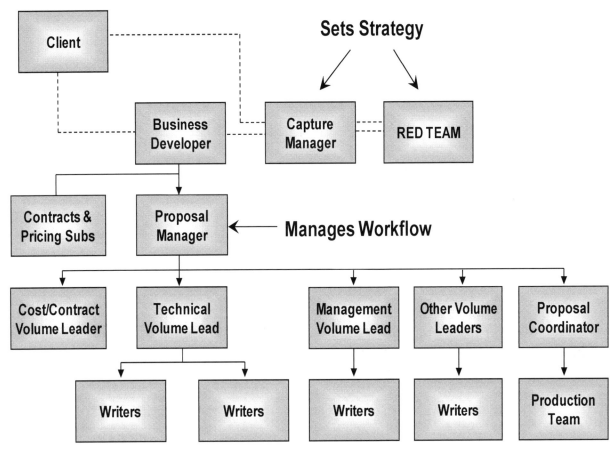

Figure 2 – Proposal Team

4.2.1   Business Development Manager

The term Business Development Manager is often confused with or used collectively with that of the Capture Manager. In reality these two functions are frequently combined, particularly in smaller businesses. In the most basic of definitions, the Business Development Manager serves to identify the company's skills and to relate them with upcoming opportunities through researching upcoming client needs versus your company's capabilities. One may view this as the highest level of marketing and planning. The Business Development Manager becomes involved very early in the process, well before a Request for Proposal (RFP) is issued.

4.2.2   Capture Manager

A Capture Manager is responsible for winning a business opportunity identified through the efforts of the Business Development Manager. Typically, the capture manager works on selected opportunities and oversees bid strategies, pricing, teaming, and proposal strategies. A major focus for the capture manager is to manage the transition from opportunity discovery to the proposal process where the opportunity is closed. The capture manager must sell the opportunity internally to the upper management staff in order to acquire the resources necessary for pursuit and proposal development. The capture manager should be an accomplished business developer who understands project management, contracts, and proposal development. The capture manager should have some familiarity with the client in order to be conversant in their terminology, policies, organizational structure, etc.

4.2.3   Proposal Manager

The Proposal Manager is the key individual charged with managing the proposal effort. This individual typically participates in marketing meetings during the pre-RFP marketing effort and should have an understanding of the technologies or programmatic issues to be addressed. In some cases, the proposed Program Manager may perform as the Proposal Manager (this is especially true on small proposal efforts or small businesses). However, if the proposal effort will involve a team comprising personnel from various organizations within your firm or from other corporations, it is best to pair an experienced Proposal Manager with the proposed Program Manager. This also applies for proposals where your firm is proposed as a subcontractor.

An experienced Proposal Manager is necessary. The novice should not be assigned as the Proposal Manager on an effort that is important to the growth of the Company. If possible, the proposal effort should be viewed as an opportunity to train at least one more Proposal Manager by using a newcomer (i.e., a novice) either as an assistant to the Proposal Manager or as a coordinator.

For you, the proposal writer, it all starts with the Proposal Manager. This individual is responsible for managing the entire proposal team's efforts to produce a responsive, clearly written, and qualified document on-time to the prospective client. The Proposal Manager typically reads all inputs from the writing team and identifies changes that he or she believes are needed. This individual is a worker, not an executive in this role. His/her job requirements are demanding.

4.2.4   Proposal Coordinator

The Proposal coordinator is extremely important to the timely production of the proposal. This individual is typically the central contact point for both the client and the proposal team. He or she receives all government modifications of the procurement and communicates this information to the proposal team. This individual performs the role of the writer's connection to the desk top publishing staff. Note that although the entire staff has access to computer usage for developing their writing assignment, there should be individuals who prepare the separate "standard copy" from writer inputs.

The Proposal Coordinator maintains the official version-oriented file of each written element comprising the draft proposal. Logistics are vital!

All writers interface with the coordinator when delivering their writings and writing updates. This individual maintains the official proposal file consisting of all written information concerning the solicitation and proposal response material. We view the Proposal Coordinator as central to the entire development effort.

4.2.5   Volume Leaders

Here we discuss the leaders of each of the volume teams. Typically, there is a volume leader for each volume required to comprise the proposal. Also note that a volume may be a major section of a physical volume. For very small proposals, this individual may be the only person assigned, thus he or she does the writing and is responsible for its response to the RFP. For most proposals, this individual leads a team of writers each assigned a section or subsection of the document under preparation. Each member of the volume team provides their writings to the volume leader for review, possible editing and/or correction, and submittal to the Proposal Coordinator for inclusion into the proposal file. Good management practices dictate that any change made to the writer's material by the volume leader is communicated and to some extent, negotiated with the writer.

4.2.6   Proposal Writers

The proposal writers assigned to the team should have the background and experience necessary to make significant contributions to the effort. Writers consist of both technical and management personnel.

4.2.7   Proposal Review Teams

Throughout the process of developing the proposal, versions of the document will undergo in-process reviews by teams of individuals outside of the writing team. These individuals comprise color-coded teams that are assigned to review progressive versions of the document under development. These teams, and more detail of the functions they perform are fully explained in Section 5.

5.   THE PROPOSAL DEVELOPMENT PROCESS

To produce quality proposals we must first standardize our proposal development efforts to create them. Beyond this, we must:

> *If you don't know where you're going, any road will get you there. Here is your destination as a proposal writer...*

➤ Develop and use a unique, meaningful theme(s) that are responsive to RFP requirements and woven throughout the proposal.

➤ Answer the question—Why Us?   More clearly, emphasize the benefits we will provide to the customer and the advantages that the customer will realize in choosing us.

➤ Where appropriate, we should provide an Executive Summary that answers the above questions. We should use the active voice in writing. We must be careful not to overwhelm the reader by presenting too much technical detail in the text. (If possible, put details in appendices.)

5.1   The Proposal Schedule

The proposal process consists of multiple phases:

1. Pre-RFP
2. RFP/Proposal Kickoff
3. Proposal Development
4. The Internal Review & Modification Process
5. Executive Review
6. Proposal Submittal

Following the submission of the proposal is the Government's review and selection process whereby they evaluate proposals and select their choices of the better of the submissions. These evaluations are predicated on a plan developed and approved by the contracting officer. The plan typically centers on technical evaluation and cost submittals. Contractors who fall below the "acceptable" line are eliminated from further evaluation. The remaining contractors enter a negotiation phase. This process may well involve requests for clarifications of your proposal by the prospective client, and a request for a Best and Final Offer (BAFO). Those firms receiving the BAFO request are in the running for award. Figure 3 depicts the development process.

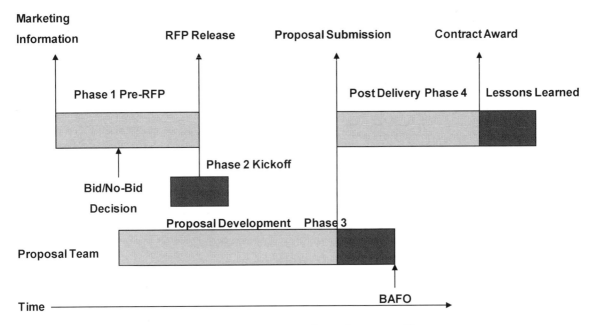

Figure 3 – The Proposal Development Process

5.1.1   Pre-RFP--Bid/No Bid Decision

As stated earlier, prior to the arrival of the RFP, your firm makes a preliminary bid/no-bid decision. This is based on marketing information and on answering a series of questions.

5.1.2   Proposal Kickoff

With release of the RFP, the writing team should be identified. A second bid/no bid decision may occur if new information within the solicitation is damaging to the company's bid position. This is typically unusual. We assume that the proposal kickoff takes place.

The kickoff meeting is attended by all members of the Proposal Team, and additional individuals with involvement. This meeting is typically presided over by the Proposal Manager, who will introduce speakers. The Marketing staff supplies customer and competitor information, win themes, discriminators, assignments, schedules, writing samples, and other valuable information that is made available to all participants. It is critical that all players have the information they need and that everyone understands his/her assignment.

5.1.3   Proposal Development

Looking ahead briefly, understand that proposal development is a sequence of study, design, writing, review, and correction. A major key to proposal methodology is the early development of the proposal outline and cross-reference matrix. The top level outline must be consistent with RFP sections C & L. Once this higher level outline is established and agreed to by the Proposal Manager and the other key players, the outline is expanded by adding lower levels to include specific requirements from other sections within the RFP. The outline is further expanded to create an annotated outline that covers lower level categories on what is required in the section. No writing will be allowed until the outline is approved.

5.1.4   The Internal Review & Modification Process

In-process reviews should be conducted throughout strategic phases of the proposal development life-cycle. A list of these reviews is shown below in Table 1.

Typically, multiple reviews of the writing team's work are scheduled and performed throughout the development process. Teams are color-coded to identify the type of review being performed. Team colors and functionality typically vary between companies. For our purpose, we will simply identify five teams. These baseline reviews may appear simple but in fact they are critical and deserve significant effort. The teams identify errors or inconsistencies of the outline of the proposal. All sections of the RFP that are required to be included in the proposal must be identified by a specific outline section. These teams will also review a list of key words or phrases taken from the RFP, and specified in writing, typically by the Proposal Manager and leadership staff. These elements will become crucial in the writing process.

Detailed functions of these teams are described in Chapter 17, Reviews.

Table 1 – Typical Proposal Reviews

Name of Review	Process
Yellow Team Review	Confirms that the outline is consistent with the RFP Sections C & L requirements. This team also reviews key words and phrases from the RFP that will be used by the writers. When complete, we never return to this phase.
Blue Team Review	Reviews expanded outline which now contains (quasi-hidden) requirements from other RFP sections such as J & M. Key words and phrases are checked again. Again, when complete, this phase of review is not repeated..
Pink Team Review/s	Reviews multiple (as necessary) versions of the draft proposal. Final approval serves as input for the Red Team. This is a review of content, not format.
Red Team Review	Reviews final version as deemed passable by the Pink Team. This is the single most critical review.
Gold Team Review	Conducts final corporate review and approval – Includes Technical and Price.

For now, consider the following for starters; typically, there are the Yellow and Blue teams. The Yellow Team reviews the initial outline created. The outline must match RFP instructions and must be augmented if necessary with additional requirements found elsewhere in the RFP.

The Blue team reviews the expanded outline, including additional RFP elements.

Following the Blue Team's approval, a final draft of the proposal is produced and provided to the Pink Team for review. The Pink Team's job is to confirm that the outline has been followed and that the written responses correlate with the RFP. The writing team corrects errors and inconsistencies as appropriate. The Pink Team may conduct multiple reviews and corrections until it deems the writing to be fully responsive to the RFP as specified by the Blue Team.

Next the Red Team reviews the updated document for technical correctness, completeness, and most important of all, that the document is convincing. This team's job is to ensure that the written material meets all of the client's stated needs and is well written. All elements of the proposal are reviewed by the Red Team. The Red Team's job is a complex one. It will perform its' review as closely as possible to that of the government evaluation team. This team is not concerned with the evaluations of the previous review teams.   It accepts the findings from the Pink team (previous review team) that the document is within the range of completeness and responsiveness. The Red Team will evaluate the quality of these features. The proposal review process is explained in more detail in Section 17 of this document.

Once the appropriate corrections are made, the finalized document is introduced to the Gold team.

5.1.5   Gold Team Executive Review

Following the approval of the Red Team, company executives typically review the proposal. Some view this as merely a courtesy. However, these executives must be satisfied that the proposal represents the company in an honest way and he/she must be aware of any corporate commitments made within the document. These individuals typically do not perform "wordsmithing", even though they are tempted and sometimes overrule this concept. The Gold Team is to review corporate commitments. As with all versions of the proposal, copies of Gold Team Review comments are saved by the Proposal Coordinator for each review version.

5.1.6   Proposal Submittal

In delivering critical proposals, the proposal should be forwarded to the Government using two independent "paths" (e.g., an overnight delivery service, Express Mail, a second delivery person taking a separate route, etc.). Reproduction costs for the extra set of volumes are low compared to the "cost" of not getting the proposal to the Government on- time.

6.   RFP DISTRIBUTION

You can't win if you don't play! However there are not enough available resources to bid on every contract: therefore, bid on the contracts where the risk/reward ratio is biased in your favor.

Remember that heavy emphasis has been placed on writing proposals for contracts where your firm has done its up-front research and preparation. Needless to say, this is not always the approach. Many firms review all RFPs received for the possibility of a bid.

When an RFP is received from the Government, it is typically routed internally to a specific individual within the firm. This individual ensures that the appropriate technical, management, business development, and contracts personnel are provided with copies immediately. The recipients then carefully read and understand the RFP, and summarize its major requirements and features on an RFP Evaluation Sheet. Evaluation forms are important because they will be used in helping to make the final bid/no-bid decision. A typical sample of an RFP Evaluation Sheet is shown in Figure 4.

Request for Proposal
Evaluation Sheet

RFP No.:

Issuing Agency:

Title:

RFP Date: Proposal Due Date:

Synopsis:

Procurement Type   Contract Type

SIC Code or Size Standard as Applicable:

Contract Length   Period of Performance

Contracting Officer   COTR or Contact:

Name: Name:

Address: Address:

Telephone Number: Telephone Number:

FAX No.:   FAX No.:

e-Mail Address   e-Mail Address

Proposal Effort Required:

Incumbent or other Competitive Assessment:

Comments/Recommendations:

_____

Signature

Figure 4 – Making the Correct Bid/No-Bid Decision

7.   PROPOSAL OUTLINE AND CROSS-REFERENCE MATRIX

As seen in Figure 5, the proper way to outline the proposal is to key the contents to one or more of the following sections of the solicitation: the Proposal Instructions, the Evaluation Criteria, or the Statement of Work (SOW).

Overriding all of this is the fact that RFP Section L, Instructions to Offerors normally dictates the format of the proposal. *You must look in Section L!*

> *Our proposal outline and cross-reference matrix will serve both as the Table of Contents and as a means by which to demonstrate compliance with the Government's RFP*

The format decision is not yours! In some solicitations, various terms and conditions found elsewhere in the RFP (e.g., in appendices) also can impact on the proposal in that they require a specific response from bidders. For the example shown, the proposal was keyed to the SOW, with appropriate references made to the Proposal Instructions and the Evaluation Criteria. Regardless of which element of the RFP drives the outline, all material requiring a response should be addressed in the cross-reference matrix so as to demonstrate total compliance with the Government's requirements.

Beyond these attributes, several other characteristics of the Table of Contents shown in Figure 5 should be noted:

The technical proposal section numbers are tied directly to the paragraph numbers in the SOW. That is, Section 1.0 corresponds to SOW paragraph C.1, Section 2.0 corresponds to SOW Section C.2, and so forth. This makes it easy for the evaluator to identify sections of our proposal and to certify compliance with the RFP's requirements.

Though not apparent (because the text of the SOW is not included here), the section and subsection headings are identical to the subjects and headings cited in the corresponding sections of the SOW.

The page numbers in the proposal are keyed directly to the proposal section numbers and hence, to the SOW paragraph numbers.

Simply stated, your outline and cross-reference matrix, which becomes the Table of Contents, should show a one-to-one correlation to all sections of the RFP that are driving the proposal.

Table of Contents				
PROPOSAL SECTION AND TITLE	Page Number	Proposal Instructions	Evaluation Criteria	SOW Paragraph
INTRODUCTION 1.1 UNDERSTANDING THE PROBLEM 1.1.1 Purpose 1.1.2 Background	1-1 1-2 1-2	L.5(b)	M.4 Inclusive	C.1
1.2 ORGANIZATION, PERSONNEL AND FACILITIES 1.2.1 Organization 1.2.2 Personnel Resources 1.2.3 Facilities	1-10 1-10 1-11 1-11			
1.3 PROPOSAL ORGANIZATION	1-11			
2.0 SYSTEM SUPPORT STUDIES	2-1	L5(a)(b)	M.4	C.2
2.1 STUDIES, REVIEWS, AND EVALUATIONS 2.1.1 Requirements Phase 2.1.2 Design Phase 2.1.3 Development Phase 2.1.4 Test/Implementation Phase	2-2 2-3 2-7 2-12 2-17	L5(b) 1	M.4.1	C.2.1
2.2 JFC INTERFACE REQUIREMENTS 2.2.1 Interface Identification 2.2.2 Interface Meetings 2.2.3 Reports	2-22 2-22 2-25 2-27	L5(b) 3	M.4.1	C.2.2
2.3 TESTING 2.3.1 Test Plans 2.3.2 Technical Approach 2.3.3 Regression Testing 2.3.4 Acceptance	2-30 2-30 2-33 2-38 2-40	L5(b) 2	M.4.2	C.2.3
2.4 DOCUMENTATION 2.4.1 Source Material 2.4.2 Format and Style 2.4.3 Resource Utilization 2.4.4 Document Version Scheme 2.4.5 Configuration Control	2-44 2-44 2-46 2-47 2-50 2-53	L5(b) 3	M.4.1	C.2.4

Figure 5 – The Table of Contents Also Includes a Compliance Matrix

The outline and cross-reference matrix is typically prepared by the Proposal Manager with support of others. Once completed, they should be approved by the appropriate Director, the proposed Program Manager, and the lead technical personnel.

The first question a review team will ask is whether or not the proposal outline satisfies the requirements of the RFP. If the format is not consistent with that required by the Proposal Instructions, the Evaluation Criteria, and the SOW, you will be directed to bring your proposal outline into conformance with these sections of the solicitation. So, if you are helping to develop a proposal's Table of Contents, save yourself and other members of the writing team from additional work by kicking off the proposal preparation effort with an approved outline and cross-reference matrix!

Finally, be aware that on some solicitations, the Proposal Instructions, the Evaluation Criteria, and the SOW may actually conflict with one another. When this happens, try to obtain clarification by submitting questions through your contracts administrator. He or she will forward the question to the Government for clarification. ***And don't vilify the RFP (that is, don't call the customer's "baby" ugly!)***. The Government's technical personnel know that the RFP may not be perfect. But they have lived with it for a long time, have fought for it throughout the agency's bureaucracy, and now are simply thankful that it's on the street. So, ask for clarification. And if none is forthcoming, meld the Proposal Instructions, the Evaluation Criteria, and the SOW together, and respond in as logical and consistent a manner as is possible.

The Table of Contents that your firm will provide, (outline and cross-reference matrix), will ensure that you have not only covered all of the topics required, but also, that the evaluator may easily find what he or she is looking for when reviewing and grading your response.

8.   STORYBOARDS

Storyboards force the writer to identify the essential information to be presented.
They take advantage of the fact that
inherently, the "scene" can be more important than the "dialogue."

Storyboards are an alternate technique to the proposal outline described earlier. Both storyboards and annotated outlines provide guidance to the writers. Storyboards are a display technique that has long been used in the motion picture and television industries. They are a convenient way to distill the essence of a presentation down to a thematic sentence, supporting sentences, and an illustration. Linked together, storyboards are to your proposal what similar collections of panels are to a planned film or television show. They provide a "roadmap" for your writers! A typical form of a storyboard is shown in Figure 6.

Once a proposal outline has been approved (and only then!), it is the Proposal Manager's responsibility, in conjunction with his key technical people, to prepare storyboard assignments down to whatever level is required (i.e., section, subsection, etc.). Then, storyboard assignments are distributed to those personnel having the technical background required to develop superior storyboards. These individuals should have also participated in developing the theme and its supporting material. .

One way to begin the preparation of storyboards is to consider the proposal theme that has been developed and to generate the picture, graph, table, or diagram that best supports this theme in the section (subsection) under consideration. Once the illustration has been drawn, you may be surprised how fast the supporting material falls into place.

Your proposal should demonstrate that you have, in effect, already started work on the job. This fact should shine through and through the storyboards you develop. Be careful of introducing new items in the proposal. Items that have not been briefed to the customer before the RFP was released imply risk. Hopefully, if the items or approaches we briefed to the customer during our marketing visits became their preferred approach, we are ahead of the game.

Whenever possible be as quantitative and as specific as possible. Instead of saying "many years of experience", say "over 10 years of applicable experience."

Use trade-offs whenever possible to identify the preferred (our) solution. This is a great way to shoot down the competition.

Use your most qualified people to develop the storyboards. *It is vital that the best technical approaches and solutions be developed quickly so that those doing the writing are provided with clear guidelines from the very beginning of the effort.*

Document / Version		Section	Author
Topic		Date	Artist
Thesis Sentence			
State the point of each paragraph:			
1.			
		Draft Drawing or Chart	
2.		**Goes Here**	
3.			
4.			
		Figure Number and Caption	
Sub-themes, AHAs, Ghosts, OH-OHs, Discriminators:			
Author:		Date:	

Figure 6 – Typical Storyboard

Once the storyboards have been prepared, they should be reviewed by the color teams previously described in the same manner as annotated outlines and draft writings.

Like draft writings, storyboards are collected and made available to the proposal team by the Proposal Coordinator. This may be easily achieved where the firm utilizes a Local Area Network (LAN). In the past, some organizations preferred to review storyboards by mounting them, in sequence, on the walls of a large room. In this case, the reviewers move around the room, make comments on the critique forms provided, and take notes that will be used in discussing the storyboards with their developers. Regardless of the techniques used, the intent of the storyboard review is two-fold:

To ensure compliance with the RFP, and to confirm that the proposal theme and technical/management presentations are viable, and that they provide a framework within which it will be possible to develop a winning proposal.

While time obviously is of the essence, it is important to understand that no text should be prepared until a storyboard has been approved. *Remember, the shorter the time frame for delivery of the proposal, the more critical is the need for detailed outlines or storyboards. You must take the time to do this!!*

Thus, if necessary, storyboard developers should be prepared to work with the review teams to revise and correct those storyboards that are found inadequate. A little time spent revising a storyboard can save days of rewrite time that would otherwise be required when deficiencies are discovered by the Pink or Red Teams.

Remember however, storyboards are not for everyone. While they are an outstanding approach to preparing the early drafts of your proposal, it takes practice and imagination to prepare them. Many firms will find it easier to "work their way" into the art of storyboard preparation through practice.

9. THEMES

Imagine having to tell the customer in only one or two sentences
why the contract should be awarded to YOUR FIRM.
What you say should basically define your proposal theme.

Your proposal theme is, of course, the central item that drives your solution to the customer's problem. Variations of the theme should be woven throughout your proposal.

The theme should go to the heart of the customer's problem. As such, it should have leverage with them, and they should recognize it immediately as the essence of the message your firm is trying to convey.

Other criteria further help to identify and manifest a theme. For example, the theme should be significant. Don't waste the Government's time by dwelling on trivia. If the theme is unique, it makes it true for us and not true for our competition. The theme should also be supportable because we will play variations of it over and over throughout our proposal.

SAMPLE THEMES

Our staff has current and extensive experience in the development of structures that are nearly identical with those required by the Army. We proudly hold letters of commendation from this (or these) client/s for the work we have performed and our adherence to the contract requirements and schedule.

The National Health Care project requires a solid understanding of the complex interrelationships of multiple Government agencies. We understand and are prepared to meet these demands through our integrated approach.

Our approach identifies and satisfies the imposition of temporary alternate methods that will ensure that the implementation of our system will cause no interruptions to current services.

Figure 7 – A Good Theme Goes to the Heart of the Customer's Problems.

10. DISCRIMINATORS AND OTHER ZINGERS

There is a way to differentiate between YOUR FIRM and the rest of the crowd: it involves using something called The D/A/G Discipline. "

There is a discipline that is sometimes used in the proposal development process called the D/A/G discipline. This process entails consideration of the following factors:

10.1 Discriminators

The "D" of the D/A/G discipline, are the ways that our concepts, approaches, techniques, organization, etc., differ from those of our competitors. For example: We are using a team composed of internal divisions. Our competitor is using a team composed of three independent corporations. Or: "Our approach to the problem is to use a microprocessor-based system. Our competitor will propose a minicomputer. Again: "Our trade-off studies value availability. His probably will value maintainability. The intent is to list as many DISCRIMINATORS as is possible so that at the very least storyboard developers and writers can use them to help the customer differentiate between your firm and your competitors. We'll also use DISCRIMINATORS in conjunction with other factors below to tell a-unique story

10.2 AHA!s

An "AHA" is an accomplishment of which we are truly proud. For example: "Our AJAX system for controlling all traffic lights throughout the city of Jacksonville, FL went from concept to initial field use in only eleven months." Or: We recently completed a complex set of Quality Assurance plans and procedures for the Health Care Finance Administration in less time than the Government allocated and at a savings of $170,000." These remarks really catch your attention, don't they? They're supposed to, because they are significant achievements to which only we can lay claim. By the way, did you notice the implied benefits in these AHAs? Benefits perceived by the reader here, would be the early introduction of needed software, and savings in both time and money, respectively. Remember, people and organizations buy benefits. Couple a benefit with an AHA, and you can deliver a real one-two punch. List all of the AHAs you can that are related to the work you are proposing to provide.

10.3 Ghost Stories (Or Just Ghosts)

This discipline is viewed by many as the equivalent of what some would call "Dirty Pool". Having been said, be aware that there are firms in your industry who subscribe to this process. In such cases, it may be said that the best defense is a good offense. If you have reason to believe that one of more of your competitors would use this approach, you may be faced with a decision as to whether or not your firm will either engage or defend. Ghosts are intended to raise issues in the back of the reader's mind regarding competitors. One should never mention the name of the competitor. Instead, we alert the evaluator a bit by identifying that problems exist here or there in the various approaches that can be used to solve his problem - approaches that may well be used by our competitors. We then identify the approach we propose to eradicate the would-be problem. Try: 'We minimize risk because our proven method solves this problem through an innovative modification of the traditional review and audit process that places emphasis on earlier error detection." Of course, this is a fictitious example, but you get the point. Or, let's say our major competitor is a team of three independent contractors while we will use personnel from our internal divisions. In this case, we might say something like: "Effective management and coordination of the field teams are the key to performing this contract within the time and cost constraints imposed. Because our personnel will be drawn from our internal divisions, which have the same management styles, accounting systems, and personnel policies, we can ensure that we will at all times be fully responsive to the Navy's requirements". See? Now the Government evaluator is thinking: "The other guy's team involves three companies. What kind of management problems will that raise? "Am I going to get the best approach for the money? Am I even going to get my job done?" List all GHOST STORIES you can think of regarding your competitors on the proposed effort.

We have to be on guard, of course, for GHOSTS others will raise about us. These "reverse GHOSTS" are called "OH-OH's", and if we don't address them, they can hurt us. So make a list of OH-OH's, and then, think about how you will counter them. For example, let's say we were two months late in delivering a data base to the Navy, and that our competitors may have mentioned this in talks with the customer. We could counter this in a manner such as: "Our firm was once two months late in delivering the XYZ data base. However, by waiting until all related Government agencies had been given the opportunity to respond to our questionnaire, we were able, at no additional cost to the client, to deliver the most comprehensive database on Greenland-based operatives that is available anywhere within the U.S. military today!" How about that? In one sentence we countered the OH-OH, delivered an AHA, and even provided two benefits (no additional cost and the best data base available). Not bad!

Now, think about linking DISCRIMINATORS, AHAs and GHOSTS. Look for things that appear on two or all three of your lists to identify the overlapping thoughts. Because they overlap, they will have the greatest impact on your readers (i.e.; the evaluators). For example, a DISCRIMINATOR by itself helps to differentiate among bidders, but coupled with an AHA!, it could really catch the reader's attention. Couple this pair with a GHOST STORY, and you have the elements that themes are made of. Whatever it is that appears on all three of your lists, it's an almost certainty that it is unique to your firm (a DISCRIMINATOR); that it is significant and at the heart of the problem (or some aspect of the problem) (an AHA!); and that it's believable for us and not believable for our competition (a GHOST STORY) because the GHOST we delivered has in some measure impeached their technical or management performance credibility.

Using "The D/A/G Discipline" is a simple approach to develop proposal themes and supporting textual and graphic materials. This concept is only marginally available if you have not researched your expected competitors.

11.   THE PROPOSAL – GETTING DOWN TO BUSINESS

Up to this point we have discussed the overall understanding of the Government RFP mandatory sections and background tasks that we as leaders and writers must perform. Now we'll take a look at the project-specific entities of the RFP. Understand that these will vary depending on the RFP under review. For instance, the Government may be soliciting proposals for any number of products and/or services. That factor applies variety to the need for specific project-related proposal responses. Within this guide, are references to commonplace sections that are required in many proposals regardless of the items that the Government is seeking to procure. This will include writing examples for some common proposal sections. We will also see references and brief descriptions of many other topics the Government may request be answered, dependent upon what the solicitation is requesting. Let's start with the Executive Summary that may be part of the cover letter that accompanies the proposal, or it may stand alone as a section.

There is only one chance to make a good first impression.

11.1   Executive Summary

Strange as it-may seem, a high percentage of a proposal reviewer's subjective opinion of proposals to be reviewed will probably be made very quickly. This is especially true in the case of high-level Government executives and military officers, who generally will look at the cover, flip through the text, and scan the Executive Summary. In that brief period, he had better see a clear summary of why the proposal goes to the list of finalists. If the reviewers form an unfavorable opinion of the proposal or if they have to search the document for answers, your proposal score may likely be downgraded.

The easiest and most effective way is to simply prepare a set of bullets that answer the question: "why should our firm be selected"? An example is given in Figure 8. This Executive Summary also conveys the underlying theme of the proposal: minimal risk through the use of an experienced contractor with proven accomplishments.

Notice too, the benefits that have accrued to this customer through his previous use of the firm's services (e.g., better solutions, immediate response, etc.). Of course, this example assumes that your firm has successfully done business with this prospective customer in the past. Nonetheless, we must help the customer find the reasons and the benefits of selecting our firm. (This type of Executive Summary is especially useful on page-limited proposals.)

When page count is not limited, a more conventional form of the Executive Summary is sometimes preferred. Here, key sentences from major sections of the technical and management volumes are excerpted in order of their appearance in these volumes to form a coherent overview of the proposal's main themes. This type of presentation should be limited to no more pages than necessary, lest it bore the reader.

It should be noted that some proposal specialists advocate writing the Executive Summary first. Using such a technique, each sentence in the Executive Summary becomes the lead or key sentence for a section or subsection of the proposal. Experience has shown that this is a difficult way to write the Executive Summary, though it does force the proposal team to do a lot of up-front thinking. Just as effective an Executive Summary will result if the thinking is first put into developing the annotated outline and, subsequently, into preparing the associated text and graphics.

SAMPLE EXECUTIVE SUMMARY
HOW YOUR FIRM'S PARTICIPATION WILL BENEFIT THE CLIENT

- We have successfully supported the Army on the predecessor contract to the effort proposed as well as on other related programs; thus we are familiar with the Army's operations, personnel, and missions.

- We have repeatedly demonstrated the ability to respond quickly to critical task requirements, and to rapidly assemble and apply the assets required for successful task performance.

- On similar contracts we have proven our ability to employ cutting edge technologies in the resolution of critical fleet problems. In our previous work, this has resulted without exception, in better solutions, earlier.

- We operate in a task order environment; we have fine-tuned our management practices to ensure that our response meets the technical requirements and is consistent with the time and cost constraints.

- Our office facilities in the greater Fort Bragg area allows us to respond immediately to all of the Army's critical requirements

Figure 8 – Make it easy on the Reader. Tell him why he should give us the job.

On large, multi-million dollar procurements, it is often worthwhile to produce a stand-alone Executive Summary that not only can be included with the proposal (say, as an appendix), but also can be slipped into the inside jacket of the binder (reviewers sometimes keep these copies). Don't worry about violating the clause that appears in Government solicitations regarding the submission of costly and extravagant proposals. You can bet that your competition is pulling out all the stops, and to the extent reasonable, so should you.

One last note: if you do a multi-page Executive Summary, be sure to use the same cover art as you did on the technical and management volumes. This will help to penetrate the reviewer's recognition of your firm and to reinforce your message in their minds.

12. TECHNICAL SECTION OR VOLUME

A Proposal is a Sales Document.

It is important to remember that when you are writing a proposal, you are selling your firm's ability to provide the government with solutions that it needs. **So write to sell!** A proposal is not a technical report you are writing for the evaluators. So, throw away the traditional style guides. Use a "proposaleeze" style guide. There are many rules for developing a good, professional proposal. Figure 9 shows ten rules, which, if followed, move your proposal up the ladder towards being selected as the winner.

Proposal writing should be straightforward and to the point. The reader should not be interrupted by the use of fancy words and complex sentence structures, or by the use of unusual or cute presentation methods. Nor should he or she be burdened with technical or mathematical details that only a specialist would find of interest. To write well, start by giving the reader an overview; put things in perspective and summarize your theme, subtheme, or topic. ***Do it up front!! Do it in the first section, first paragraph , first sentence, etc.*** Some evaluators only read the start of anything. This is where you put what you want them to remember above all else. The next sentence (paragraph, subsection, section, volume, etc.) should expand on what came before it.

The first and last paragraphs of a section, and the first and last sentences of a paragraph, are the big ones. This is because you generally introduce your thought with one and summarize it with the other. And, the last sentence can also be used as a transition to the next paragraph.

Proposal writing is just like any other form of writing. There are various styles, voices, and structures available to you. However, most winning proposals generally are prepared using the guidelines presented in Figure 9. Examples of text from several winning proposals are shown in Section 13. After reading them, it should be apparent that "easy reading is hard writing."

12.1   Writing Guidelines

At this point your proposal outline is consistent with that required by the RFP. Your writing team is well aware of the theme/s that has been developed. It is time to put your team's plan into action. Before starting, THINK!! You want your proposal to be responsive and to present your firm and its solution to RFP requirements in the best manner you possibly can. Figure 9 will help in deciding how to present your story.

WRITING STRATEGIES

- **Themes -** Ensure your themes are presented in more than one section of the proposal, not simply in the Technical Approach. Sections where I propose that you place the themes include the Management Plan, any Quality Assurance Plan/Approach and yes, even resumes. Get your message across early and often.

- **Benefits –** The client wants and needs beneficial solutions. Emphasize how your products and/or services benefits the client, not simply satisfying the requirement.

- **Be the Client –** Write in a knowledgeable manner to ensure that the client wants your firm to be on his or her team. Reference visits to the client's facility and speak convincingly in presenting your knowledge of their needs. Be specific. Speak the client's technical language.

- **Write Like You Speak –** Use plain convincing language. Don't force the Government's evaluator to reach for a dictionary.

- **Use Illustrations –** Figures and Tables get your message across better than words. These make your response easier to understand, and shows the client that your approach is logical. Be aware that these elements need to be well thought out.

- **Justify your Firm's Capabilities –** Proudly mention identical or near-identical work your firm has performed in circumstances similarly related to the subject of which you are writing. Let the client know that "We've done this before."

- **Write in an active Voice –** Do not be passive. Emphasis should be on how your firm "will", not "can".

Figure 9 – Rules for Writing Proposals That Win.

12.2   Writing Samples

Remember to be active and concise in your writings. Show the client that your firm understands and has the solution to his or her needs.

Active:

XYZ's task order management approach is tailor-made for the multiple task order environment of the NAVSEA contract. We have applied this methodology on numerous task order contracts, and we have continually improved it through those experiences. We will provide the Project Director with access to corporate resources, administrative support, and tools to support task order procedures that are taught in our management training program. Our managers have been thoroughly trained and are heavily experienced in the approach to managing multiple task orders.

Concise:

XYZ's corporate standard methodology for new system development is John Smith's Secure Partition Engineering (SPE) Methodology, as documented in the attached material. SPE is an architecture-based, business-oriented approach to developing secure partitions. It supports organizational goals and objectives implemented in a standards-based environment to ensure security

12.3   Win Themes and Discriminators

12.3.1   Example 1

XYZ recognizes the importance that the United States Courts places on the integrity of the data of the Federal Judiciary. To provide information security and to protect the privacy of individuals, XYZ will immediately implement a Security Management Plan (SMP) as part of the overall Contract Management Plan (CMP). The SMP shall be written using guidance provided in Federal law, agency policies, regulations, and guidelines. This SMP will aid the XYZ developers, system analysts, and technicians in understanding and meeting the security requirements throughout the life of the contract. Providing security guidance shall be the stated responsibility of the Project Director, and the protection of the information entrusted shall be the responsibility of each XYZ employee.

Past success is the best predictor of future performance. XYZ's 6 years of experience concentrated in systems analysis and programming support services for Government agencies ensures a proven low-risk partner.

12.3.2   Example 2

XYZ's Quality Control and Testing (QC&T) activities focus on finding errors. It is conducted through each phase of the life cycle, thereby ensuring the testability and accuracy of stated requirements, and the traceability and compliance to stated software requirements. Furthermore, testing ensures that software and documentation satisfy system requirements and objectives. In all cases, QC&T activities are controlled by customer requirements, definitions, and standards.

QC&T capabilities were demonstrated on XYZ's Department of Defense (DOD) contract, where XYZ implemented QC&T through process review audits, peer reviews, and application test plan development and execution. XYZ successfully applied the IEEE standard for software quality assurance and verification and validation plans in conjunction with the SEI CMM.

13.   SAMPLE PROPOSAL SECTIONS

Depending on the Government's RFP, the technical volume (or section), will consist of multiple sub-sections depending on the client's requirements. Below, the reader will find multiple fictitious project references that may be required. These are samples of partial writings, and are provided simply to provide the reader with insight to proposal writing and formatting. When preparing these sections for your own assignment, remember key words and phrases. Capitalize them, make them bold, prepare in a different color, or whatever your team's standard becomes.

13.1   Past Performance

The following material, while based on sections of past winning proposals, is simply fictitious in terms of people, the firm, and Government agency.

Introduction

At Federal Solutions Inc. (FSI), our goal is to provide the Government with high quality solutions to all required legal, technical and operational requirements. We at FSI provide a complete range of **legal support representation** to protect our government from erroneous declarations and judgments caused by conflicting legislative proposals.

Having worked successfully to help the clean-up of poisonous materials from rocket misfires, to technical assistance for Homeland Security personnel, our action teams have left positive marks on multiple Government agencies through a variety of functions ranging from security operations support to support of mission-oriented assignments for our military forces. **FSI's performance on its recent NASA contract is very closely related to the requirements of this solicitation as can be seen in the following.**

FSI performed **administrative and logistics** services for the Technology Application Office (TAO) conforming to the procedures and polices specified in AR 710-2, AR 735-5, DA Pam 710-21-1, AR 750-1, and SPBS-R (TDA). **Logistics services** include: maintaining administrative files; requesting, receiving, inventory, storing, transferring, loaning, issuing and maintaining, supplies and/or equipment; acquiring maintenance support; training personnel; property accounting; and providing routine and specialized support as requested by the client.

FSI reconciled property authorizations as prescribed by AR 71-31 and AR 710-2, DA Form 4840-R (Request for Type Classification Exemption/LIN for Commercial Equipment), and DA Form 4610-R

(Equipment Change). Documents that support entries to the property listings were marked "posted" and dated/initialed and posted to the Document Registers, supporting Document Files and Property Books.

FSI accepted supplies and/or equipment turn-ins and prepared DA Form 3161, DA Form 2765 or other allowable form. FSI reported excess automation equipment to **Defense Automation Resource Management System (DARMS)** for disposition using the current approved method. FSI also prepared appropriate turn-in/transfer documents, posted the transaction to the applicable Document Register, Property Book, and document files. FSI also accepted supplies/equipment for turn-in at a designated turn-in point. All supplies/equipment were returned to the Supply Support Activity after determination that the supplies are excess. FSI ensured that serial, lot or registration numbers were annotated on transaction documents to assure **serial number accountability.**

Customer Point-of-Contact Information:

Name:	John Wilson
Address:	Technology and Supply Office
ATTN:	John Wilson 6272 Wonderland Road Fort Apache, AZ
Telephone:	(123) 456-7868

Ensure that you **highlight** key words or phrases from the RFP. It aids the evaluator!!

13.2 TRANSITION PLAN - Fictitious Sample Proposal Section

The following material, while based on sections of past winning proposals, is fictitious in terms of people, the firm, and Government agency.

Federal Solutions Inc. (FSI) is fully prepared to conduct a smooth, seamless phase-in with no disruption to ongoing activities at the U.S. Veterans Administration (VA). In so doing we will apply our extensive transition and operational experience to the individual needs and requirements of this contract.

Of these two transition activities, the most immediate is phase-in. FSI is fully prepared to conduct the phase-in in a smooth, professional manner.

Our top two priorities are:

> ➢ Recruitment of the "best and brightest" of the incumbent workforce.
>
> ➢ Establishment of rapport with key VA Unit personnel

Please note that FSI has extensive experience with phase-in efforts similar to those required for VA. **These experiences include NASA under contract ABC123, and with the Department of Homeland Security under contract 123ABC.** As described below, we have developed a highly successful process for this function.

Fox example, we staffed our operation at **Department of Homeland Security** with 44 personnel in a 13-day transition period. We understand the risks, especially the possibility of failing to provide adequate staffing, and our procedures prevent these risks from becoming reality.

The FSI Phase-In Team will consist of individuals with extensive contract phase-in experience. Table 2 depicts their primary responsibilities.

Table 2 – Phase-In Transition Team

Transition Team Position	Responsible Individual	Primary Responsibilities	Interfaces with
Phase-In Manager	FSI Project Manager	Primary interface with VA on all transition activities	VA Contracting Officer and COTR. FSI Transition Team
Assistant Phase-In Manager	Chief Computer Operations Analyst	Interview and hire desired incumbent personnel in Washington	COTR and incumbent personnel.
Technical Analyst	FSI Project Manager	Interview and hire selected existing staff	FSI Project Manager and VA personnel
Financial and Contract Managers	FSI Corporate Office personnel	Review all project financial and contractual data requirements and integrate into FSI financial and contract management systems	VA Contracting Officer and finance officials
IT Representative	FSI IT personnel	Set up internal computer network	Transition Team
Logistics Manager	FSI Corporate logistic personnel	Complete property inventory	Transition Team
Human Resources Manager	FSI Corporate HR Manager	Refine Recruiting Plan Oversee HR activities Support as required	Transition Team
Human Resources Recruiters	FSI HR Generalists	Travel to Arlington to coordinate hiring of selected personnel	Incumbent Employees or FSI proposed personnel
Quality Assurance	FSI Corporate Office personnel	Development of Project Quality Plan	Project Manager

The anticipated number of hours per week required of each position to complete their assigned phase-in task is shown in Table 3.

Table 3 – Estimate of Required Phase-In Hours

Position	Week 1	Week 2	Week 3	Week 4
Phase-in Manager	30	10	5	10
Asst. Phase-in Manager	10	5	5	10
Technical Analyst	5	5	5	5
Finance Rep	20	10	5	5
IT Rep	20	20	20	20
Logistics Manager	0	40	10	5
HR Manager	40	40	20	10
HR Recruiters	40	40	20	10
Quality Assurance	40	20	10	10

Transition Actions

Again, FSI is experienced at transitioning work and workforces after contract award. For the VA contract, as with others, we will ensure that all phase-in actions are complete in time to assume full responsibility within the 30 days allowed.

Table 4 identifies our major phase-in actions. The actions shown in the table (and others) are discussed in more detail in the remaining sections of this plan.

Recruiting and Hiring Incumbent Employees

Recruiting and hiring incumbent personnel during a contract transition is a critical and sensitive challenge. Our primary objective during phase-in will be to recruit all of the qualified incumbent personnel who have performed well and wish to remain with the program. The only exception will be the incumbent Project Manager. We feel it is essential that this position be filled by a FSI individual.

Table 4 – Major Phase-In Actions

Action Item	Impact/Benefit
Acquire existing qualified staff	Ensures qualified core staffing
Acquire any required new hires	Completes the staffing of all positions
Conduct/attend program familiarization training	New key personnel and employees are made familiar with site operations
Acquire contract documentation (procedures, regulations, policies)	Provides FSI with detail of actual operations
Introduce Government staff to new employees	Ensure rapport and good communications
Conduct thorough inventories of equipment and supplies	Ensures availability and accountability of all required equipment and supplies

Recruiting for Key Positions.

FSI has completed its recruiting for key personnel. Five of the seven positions requiring resumes will be filled with individuals who are currently employees of the incumbent contractor. Utilizing these individuals will help ensure a smooth transition and retention of program-specific 'historical' knowledge.

Ensuring Success through Effective Communications.

FSI has conducted successful incumbent recruiting efforts on multiple contracts, and we have done so in the face of situations ranging from friendly, to adverse, to circumstances where incumbents were no longer providing service or willing or able to participate. These experiences taught us that prompt and frequent communications with incumbent personnel is the key to a successful transition, and we intend to apply those lessons in this phase-in.

Recruiting Methods.

Immediately after contract award, FSI will host an after-duty hours, off-site open house meeting for the incumbent workforce at the Arlington location. The purpose of these meetings is two-fold. The first objective is to acquaint incumbent employees with our company, our background and our intentions for the workforce. The second is to hand out recruitment materials and make appointments for interviews. We will provide the incumbent employees with FSI company brochures, a briefing by our Project Manager on our approach to meeting labor requirements (recruiting most of our personnel needs from the incumbent workforce), hand out application packages and make appointments for interviews.

Contingency Recruitment Plan.

If we are unable to fill all of the staffing needs with incumbent personnel plus our identified key personnel, we will quickly hire 'from the outside." In this regard, we have numerous applications on file that we can draw from, and we have a professional HR recruiting staff that will be prepared for this eventuality.

Contingency Training Plan.

If hiring new personnel does become necessary, the new hires will require training so that they may quickly function in accordance with the program's existing guidelines and procedures. If this need occurs, we will meet it primarily by utilizing the individuals we hire from the existing staff. Any shortages of trainers from this source will be met by incumbents that don't join us, given that the incumbent contractor is tasked to provide both phase-out and phase-in support.

Training

Again, we will provide orientation and training for any employees that are new to the VA program. And within the 30-day transition period, each of our employees will have been certified at the level of professional requirements required by the contract.

Training will be conducted primarily during off-hours by incumbent personnel who agree to join FSI. Training will consist primarily of familiarization with the processes and procedures that are specific to the VA project. Trainees will also be provided with materials for self-study.

As a potential supplement, we will take advantage of any training required of the incumbent contractor as part of their Phase-Out Plan.

Inventory Issues

FSI understands and is familiar with recording, accounting for and protecting inventory items provided by the Government for use during contract performance. During the contract phase-in period FSI accounting personnel will verify the Government's inventory identified in RFP Section J.5, Government Furnished Property. The inventory will be recorded within the FSI Contract files as required. Any medicinal inventory that may become the responsibility of FSI will also be identified and stored in the contract file.

Phase-In and Phase-Out Communications with VA

FSI appreciates the need for up to date, close and thorough communications with VA during the phase-in period. We will maintain daily contact with Government transition personnel. These contacts will consist of informal quick reference, quick response discussions. On a weekly and more formal basis, our Project Manager will collect and provide written progress and problem documentation to the appropriate Government personnel. These documents will consist of progress made vs. progress planned indicators. They will include but not necessarily be limited to the following.

➢ Staffing progress

➢ Orientation and training progress

➢ Inventory progress

➢ Records, files and documentation transfer progress

➢ Problem identification and resolution information

➢ Planned vs. actual phase-in progress statistics

Based on the above plans, we are confident that we can complete the 30-day phase-in period successfully, with no interference to ongoing tasks.

13.3 On-Going Recruiting and Staffing

The following material, while based on sections of past winning proposals, is simply fictitious in terms of people, the firm, and Government agency.

Throughout the performance of this contract FSI will maintain its staffing operation as required. We offer employment opportunities only to those individuals who fully satisfy the requirements of the position and who appear after investigation to meet our high standards of professionalism. .

Recruiting Plan.

Our recruiting process has three basic components. First, we always seek to promote from within. If this is not possible, we move to our database of past applicants. Finally, as shown in the diagram, we exercise a wide range of recruiting options. Full candidate screening is conducted internally, and customer approval is sought prior to placing any successful candidate in our client's facility.

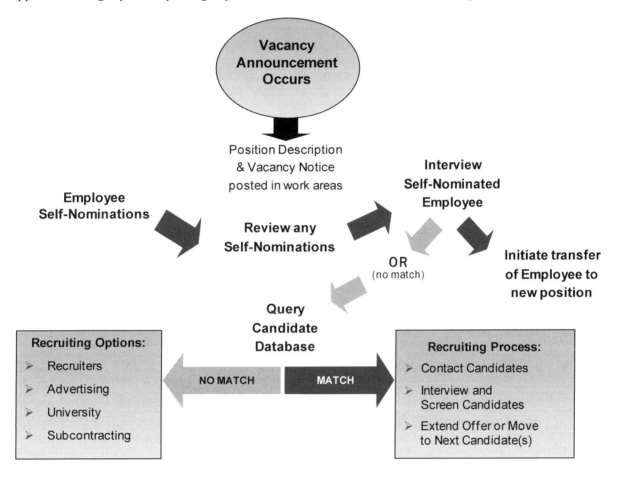

In today's market, attracting and retaining highly qualified personnel is a challenge for all companies. At FSI, we consider ourselves fortunate that we have been successful in outpacing our competitors in this arena. We believe that our success is attributable to the quality and desirability of our Employee Compensation Plan and the way we treat our people. We have learned through our experience that being pro-active in our concern for our employee's well-being gives us a decided advantage over our competitors. That of course, results in higher success for our company. We know that qualified personnel tend to want to work in an organization where their efforts are recognized as substantial contributions to the success of the business. We provide an environment for this success by developing innovative plans for improvement of our clients' operation, and by making our professional men and women part of the solution, not part of the problem.

FSI Staffing Activities

FSI uses a number of approaches to successfully staff. Consistent with our commitment to attracting highly qualified staff, we use an experienced, dedicated full time recruiter at the Director level. The ranges of approaches we use include:

- Partnership with other companies for subcontracting
- Newspaper Professional Help Wanted Ads
- Hosting and Attending Professional Career Fairs and Trade Shows
- Employment Agencies
- Employee Referral Program
- Internet and Web Site Recruiting
- Recruiting at Universities and Colleges
- Targeted Direct Mail
- Tele-recruiting
- Dedicated "on-site" recruiters utilized, when warranted, in addition to our centralized recruiting functions

Some of the approaches we use are described briefly below.

Career Fair and Trade Shows:

As a leader in our industry, FSI Services participates in a variety of career fair and trade shows held in the areas where we perform. In support of these functions we advertise in leading newspapers in the area local to the function. We use hospitality rooms to meet with and discuss opportunities with prospective employees who are in attendance.

We have committed to repeat this process periodically as required. This enables us to identify and attract qualified individuals for our client's requirements. The details of our approach are identified in the following paragraphs.

Newspaper Ads:

FSI Services places professional help wanted ads in leading newspapers to support our staffing requirements. As stated above, we also place these ads in support of FSI Career Fair and industry Trade Shows. These are designed to attract qualified personnel for current vacancies plus the resume stream necessary to keep the candidate pipeline for future vacancies in flow.

Recruiting Firms:

Another significant aspect of our recruiting approach is the selective use of recruiting agencies. These dedicated "teammates" continuously identify qualified personnel to meet FSI's requirements. Typically, the firms will compete for the identical staffing requirements. FSI will select the best candidates for assignment to its contracts. Our goal is to establish a probation period where the employee who is provided by the recruiting firm is on a trial period. After a brief period of time, FSI will make a determination regarding the permanent hire of these individuals. The best of them will then become FSI employees. FSI retains the right to dismiss any personnel from the recruiting agency for performance reasons. This "try before you buy" concept has proven to be of substantial benefit to our client's and FSI.

Internet and Web Site Recruiting:

Over the past several years, E-Recruiting has made a major positive impact on recruiting. Access to resumes through national databases is instantaneous. This has revolutionized the process of identifying employee candidates in nearly all work disciplines and in all geographic areas.

FSI Services hosts its own web site at www.FSI-services.com. Prospective employees can identify job opportunities through visiting our site and communicating with our staffing department electronically. We also use nationwide Internet recruiting services, including a strategic partnership with Recruiting.Com.

The Interview and Screening Process

The highest priority for candidate sources is our Self-Nomination Program. This is followed by our Employee Referral Program. The third priority for candidate sources consists of the remaining sources mentioned above. The Director for recruiting, assisted by the Human Resources Department will screen all candidate resumes. Resumes reflecting individuals who appear to meet the position requirements will be delivered to the hiring manager or director. The hiring manager or director will interview all such candidates. All candidates other than self-nominating candidates are required to provide written permission allowing FSI Services to conduct reference checks to verify education, employment history, and any malpractice proceedings. These checks are done only for candidates who are targeted to receive an offer of employment from FSI Services. Additionally, candidates may be interviewed by our clients for acceptability.

Self Nomination Program

A very significant element of our Recruiting and Retention plan is our Self-Nomination Program. FSI personnel have the opportunity to nominate themselves for positions that are attractive to them. When new staffing requirements become available, they are identified by a posting system whereby the positions and related particulars are posted in all of our work areas. These positions will remain posted for a one-week period to allow the self-nomination process.

For Self-nominations where the nominee meets the position requirements, an interview of the candidate is scheduled. This occurs prior to any outside candidate being considered. The Director of the Program who will lose the employee is notified and replacement actions are initiated. Negotiations are conducted between the two programs to reconcile conflicting needs. If the employee is critical to his present assignment, he or she will not move to the new project until a suitable replacement is found.

The Self-Nomination program is a major segment of FSI's career advancement initiative in support of its employees.

Employee Referral Program

We recognize the extreme value of our employee's judgment in helping to determine who will become their co-workers. Employees typically recommend only those candidates who in their judgment would be assets to the company. They know that they are strengthening the competitive hand of the company by recommending colleagues who will make significant professional contributions. We reward the employees financially for this service because we know that the probability of outstanding performance by these candidates is higher than through any other recruiting channel.

Employee Retention plan

FSI enjoys a staff turnover rate that is less than the typical experience of our industry. We have had outstanding success hiring the right personnel mixes, giving FSI the ability to provide superior talent to meet our client's staffing requirements.

Employee incentive programs, employee recognition awards, and other financial rewards for superior performance make FSI a company that retains its employees. Added to these are very competitive benefits—a matching 401K program, excellent health and dental packages, tuition assistance, and other full-life benefits that make FSI the company of choice for many highly qualified professionals.

Pay For Stay

FSI recognizes the value in having its key personnel stay on the job until the job is completed. While there are times when this is not possible, we do propose extraordinary steps to provide the right incentives for the employee to stay. Foremost among these, at the discretion of the company, are a cash bonus to key personnel who remain on the job to completion. As a privately owned business, we can make these decisions.

Substitution of Personnel

Recognizing that even extraordinary actions such as Pay for Stay do not always achieve the desired result, we must always look to the possibility of personnel replacement. FSI will also implement an "understudy" approach where we identify on-board qualified personnel as potential replacements for our key employees. The understudy personnel will benefit from a system of cross information where they will remain cognizant of their potential roles as replacement personnel.

When it becomes necessary to replace personnel with new hires, we make every effort to ensure that we do not assign new hires to difficult tasks where they will be in a "sink or swim" mode. Instead, we put forth the effort to replace personnel on small tasks with on-board FSI personnel who have a basis of understanding of the work environment so that we may minimize the learning curve. We place new replacement hires on larger tasks where they have the benefit of "learning the ropes" from existing project staff.

As can be seen, FSI leaves no stone unturned to satisfy our client's personnel needs.

14. A MANAGEMENT SECTION OR VOLUME

The content of the management volume is of vital importance to the win strategy.

The following material, while based on a section of a past winning proposal, is simply fictitious in terms of people, the firm, and Government agency.

As a Government contractor, you firm is in a business that can be classified as the provision of professional products or services in an environment that is often characterized by the performance of simultaneous tasks. The ability of a contractor to manage the assigned contract work is of great interest to the government. As a result the government often requests descriptions of various processes with your firm's repertoire.

Government contractors should create a "boilerplate" library for purposes of expediting the proposal development process. These generic sections may be used as baselines from which to produce tailored material for the prospective customer.

With the above as background, it may be likely that the prospective Government client requires detailed information relative to the following contract management related areas:

Organization	Plan of Operation (this contract) Task Order Processing
• Program Organization (chart and description, generic and contract specific) • Relationship of the firm's Program Organization to Customer's Organization • Management Authority (generic and contract specific) • Management Responsibility (generic and contract specific) • Core Personnel	• Task Plan Development • Contract Administration/Coordination with Contracts • Coordination with the Government • Scheduling (contract specific) • Phasing of Tasks • Resource Loading • Peak Loading Contingencies • Critical Milestones • Reviews • Personnel (contract specific) • Selection of Personnel • Key Personnel • Labor Categories • Resumes (Tailored to RFP) • Related Corporate and Personnel Experience • Contract Information on Efforts Similar to that of the Work Described in the RFP • Government COTRs (including Address and Telephone Number, related to contract information noted above) • Relationship of Previous Work to SOW of RFP • Significant Accomplishments on previous, related Efforts
Facilities	
• Locations (Including Proximity to the Government's Program Office) • Descriptions (Square Footage, Conference Rooms, etc.) • Computer Equipment • Reprographic Equipment • Proximity to Public Transportation • Security-Related Matters • Subcontractors (if any) • Subcontractor Plan • Communication and Reporting Policies	

The development of the management volume (or section) should be approached in much the same way as is the technical volume. That is, outline and cross-reference the appropriate parts of the RFP, develop themes, prepare storyboards, etc. And while the past proposal library contains substantial management proposal material from past successful proposals that is to be used as sources of information and data, keep this in mind: the management volume of the proposal must be tailored to customer requirements and benefits. Following is a brief sample of a proposal section on workplace management for a fictitious government contract.

14.1   Program and Task Order Management

Federal Solutions Inc. (FSI) is fully capable of managing all aspects of services contracts. During our corporate experience of more than 7 years, we have acquired and developed the knowledge, skills, and management procedures to effectively and efficiently ensure that the tasks are accomplished within the schedule and costs. Chapter 15 presents a detailed example of a Task Order Management Plan write-up that has been suitable in multiple proposals.

14.1.1   Program Management

The management and reporting for the Bureau of Public Debt contract will be carried out from within our successful organization structure. Our designated Program Manager, William (Bill) Sanders is thoroughly familiar with this solicitation and he is committed to ensure that FSI gets off to a successful start. As many in your office know, Bill Sanders has substantial Public Debt experience. Initially, he will spearhead the contract start-up activities and the recruiting process. He will be supported by a Task Manager (TM) for every task order. For one-person tasks, the individual assigned to the work will also perform TM functions.

We recognize that our contract will consist of a variety of task types. Some will be single service tasks where we are required to deliver personnel with specific skills to perform a simple one-person statement of work. Other tasks will require planning and the designation of multiple individuals with varying skills, all necessary to achieve the task objectives. While we do not apply management overkill to simple tasks, it is important to note that although we scale the details of our management approach to the task at hand, the same basic management functions must be performed for all tasks. This approach has been proven over our years of government support.

15.   TASK ORDER PLANNING

This chapter discusses the operation of Task Order contracts. Your written proposal may well require these processes. Using the fictitious firm of FSI again, your firm's process should contain dialog similar to the following. A key FSI business principle is centralized planning and decentralized execution - planning for our client program as a whole, and then executing individual task orders as standalone projects. Our system of centralized project planning and subsequent monitoring and control of decentralized execution is a dynamic and continuously iterative process. It provides the structure and flexibility to incorporate new tasking and reallocate resources as necessary for the most cost-efficient customer service and task order performance possible.

Project Management Plan (PMP).

We develop Project Management Plans for each of our contracts. The complete PMP will address the planning, organization, staffing, and control of multiple tasks and resources to accomplish our customer-defined objectives, at defined quality levels, within time and cost constraints. At the start of the contract, the FSI management team will have a draft PMP in place as an initial baseline. We will then meet with the designated Government personnel to begin refining this plan. Following Government guidelines, FSI will create a master document detailing the scope and direction of the program, which will be reviewed and revised as the program progresses. The PMP is viewed as a "living document".

One of the most important objectives of the PMP is to develop a program-wide staffing plan that will allow us to proactively address task staffing ramp-ups and ramp-downs, and allow for orderly transitions of staff from task to task. Transition staffing, where FSI will offer employment opportunities to outgoing contractor personnel will also be addressed. The PMP will also address anticipated staff training needs. These needs will be translated into a staff training plan.

Work Plans.

Work Plans at the task order level provide an ordered framework for planning and controlling the work necessary to meet task-specific technical objectives, respond to change, and monitor cost, schedule, and technical performance.

The plan presents the relationship of the specified subtasks in the Task Request statement of work (SOW). It also includes a Work Breakdown Structure (WBS) expanded to show the relationship of all elements supporting the task order and provides a sound basis for cost and schedule control. By decomposing the entire effort into successively smaller entities, we ensure that all required products are identified and addressed. Our Work Plan is a key tool in preparing Task Order proposals in response to government task orders.

Our Program Manager will implement the task order handling procedures during the initial contract period. Since the procurement will have multiple awards, we assume there will be a post-award conference where successful contractors will receive instructions on the preferred task order processes. We may at that time also receive SOWs, and solicit information on pending tasks to determine information needed to specify task order staffing requirements, workload estimates, and technology needs. Thereafter we will receive task requests electronically or by facsimile or mail from the issuing government office. We will prepare task order proposals and submit them in hard copy and/or electronically to the CO or responsible COTR within agreed upon times after the issue date. FSI Task Managers (TM), who are working supervisors, with overall guidance from the Program Manager, will be fully responsible for task order preparation, task, negotiation, staffing, initiation, tracking, and close out.

Receiving and Evaluating Task Orders.

FSI will receive task order requests from the ordering office. These task orders will be reviewed by FSI and used as the basis of our Task Order Proposal.

Estimating Task Order Proposals.

Tasks will be estimated using the Delphi method of professional estimates using judgment and experience and use of automated spreadsheets using macros and formulas in Microsoft Excel. These sophisticated spreadsheets capture resource selection, resource estimates, and interdependencies within the detailed Work Breakdown Structure (WBS). They can be linked to Microsoft Project to produce Gantt schedules, as well.

Preparing Task Order Proposals.

FSI TMs, assisted by knowledgeable technical staff, will carefully prepare all task order proposals. The completed task order proposal will present a clear, concise, accurate, comprehensive, and easy-to-negotiate task solution. Each proposal will certify adherence to contract staffing guidelines. Each request will be assigned to a TM. The exact nature of the task request will determine the assignment. The TM will evaluate the task order to determine if a pre-proposal meeting is needed, and then select a team to meet with the government client to better understand the requirement. Specific data elements in each task proposal will include:

- Subtasks, activities, and functions that will be performed to complete task requirements, including baseline metrics
- Task Management and Reporting
- Assumptions used (if any) to develop proposed solution (complex tasks)
- The beginning and ending date, including intermediate milestones and acceptance criteria
- Quality Assurance Measures
- Technical approach including any productivity tools and techniques to be applied
- Proposed Labor Categories and assigned duties
- Resumes if and as required

Figure 10, Task Order Proposal Development Process, shows our proposed task order preparation process. These repeatable processes enable us to control all task requests; produce accurate and consistent proposals that clearly demonstrate a comprehensive understanding of task requirements; support timely negotiations; and ensure all changes are in accordance with contract terms and conditions.

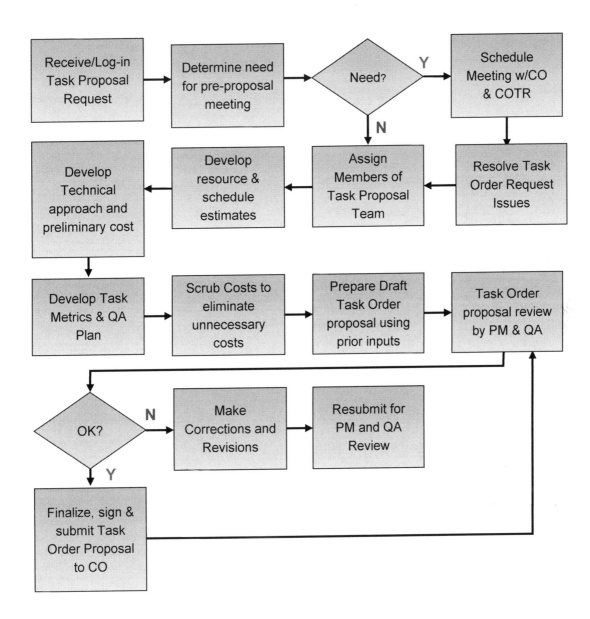

Figure 10 – Task Order Development Process

Reviewing and Submitting Task Order Proposals.

The Program Manager or his designate and a Quality Assurance representative will review all proposals to confirm the reasonableness and consistency of the estimate, ensure compliance with contract terms, validate the technical approach; and ensure we are presenting the best value solution. After review, the proposal is signed by the Program Manager or TM and delivered to the Contracting Officer.

Negotiating Task Order Proposals.

The Program Manager is fully authorized to negotiate all task orders, commit FSI to proposal terms, and reach negotiated agreements in a single session. If changes are needed to the proposal text or schedules, negotiated changes will be made and forwarded to the ordering office in the form of change pages.

Task Order Tracking.

FSI's task order tracking and oversight procedures enable our managers to track actual progress on the task order compared to the plan presented in our proposal, as negotiated. As shown in Figure 11, Task Order Control Resources, our procedures include periodic reviews where we evaluate our performance against the plan we have put into place with government sanctions.

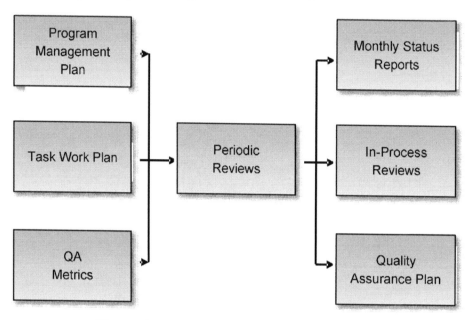

Figure 11 – Task Order Control Resources

Establishing the Oversight Team, Resources, and Tools.

Our Program Manager will initiate and execute tracking and oversight activities for each government task order. He or She will assign a team including him/herself, the task leader, the responsible TM, and a representative of the Quality Assurance team. This team will initiate use of standardized task order tracking templates within the project.

Establishing Tracking Procedures.

Using these standard task order tracking templates, the tracking and oversight team will define tracking indicators for the task based on the metrics established in the original estimate, including specifications for reporting. The team will schedule In Process Reviews (IPRs), walkthroughs, and events associated with project milestones and deliverables specific to the task requirements.

Tracking Progress on the Task Order.

To track progress on the task, we will analyze and document results from our performance, particularly focusing on any deviations from the baseline measures established for the task (as negotiated). As we perform, we will take additional measures on an ongoing basis and compare them with the baseline measures.

Task Order Controlling.

FSI's control procedures enable our managers to monitor actual progress on the task compared to the plan presented on our task order proposal, as negotiated. As shown in Figure 11, our defined, structured processes include the conduct of formal progress reviews using information from multiple sources.

In addition to performing ongoing tracking of task order indicators, we will hold regularly scheduled reviews both internally and with client management. These reviews will include IPRs, walkthroughs, and other meetings convened to communicate project performance, deviations from plan, corrective actions taken, and results of corrective actions.

We will follow a standard IPR format that provides an overview of the task order project, a discussion of issues of concern to the task leader, graphical display of cost and schedule status relative to the proposed plan, and identification of risk factors in critical areas of the project. For each risk area on complex tasks that is not shown as optimal for task order success, our task leader will develop and present a risk management strategy outlining how we plan to mitigate the risk.

Personnel Transitions from Task to Task

FSI will establish a core staff of knowledgeable, customer experienced staff to provide continuity. FSI's management planning and scheduling process, continuous communication and coordination between FSI managers and their government peers, and establishment of a core technical team will ensure smooth transitions from one task to another. FSI is very experienced in the management and transition within task order contracts. Our objective is to provide seamless technical support to the customer requirements without having to concern our client with the administrative logistics of multiple task orders. That is our expertise and we recognize it as our job.

Management and Scheduling Process

The Project Management Plan consolidates many pieces from multiple task orders to provide overall guidance, vision, and management of the work to be performed. As current task orders near completion, the TMs will examine the staffing skills, and compare them to the planned upcoming tasks. Any mismatches will be quickly identified. Resumes of staffing for which there are no new tasks are referred to our Human Resources Department (HRD); notification of requirements for which there are no staffing are also sent to HRD. Using the PMP, FSI will produce seamless transitions from tasks nearing completion to new task orders in the initial stages of implementation.

Continuous Communication and Coordination.

An extremely important part of this process is communication. FSI will listen to the customer and aggressively plan for task-to-task transitions. By enhancing communications across the customer/contractor interface FSI will have enhanced knowledge concerning upcoming tasks. This will also enable us to effectively plan for and execute task-to-task transitions.

Core Technical Team.

An important and necessary aspect of FSI's approach will be to establish, based on anticipated average workload, a core of knowledgeable, customer experienced staff. We call it "critical mass management". Members of this core group will be the first source of staff for all active task orders, as their skills are needed and are available. This enables staffing across multiple, usually overlapping task orders with the most knowledgeable customer specific experienced staff. It enables overlapping task orders to start and end as they should, with continuity to the entire program. Second, it keeps a core staff "on board" and available to support customer Task Orders. This aids staff retention and encourages teamwork and camaraderie in starting up and performing the task. This establishes a far more efficient task order performance, which in turn results in less cost to customer and produces higher-quality products and services.

16.   RESUMES

Resumes of personnel working within the firm are a vital commodity in government contracting, particularly when services are to be provided. Resumes may also be important in proposals that offer equipment sales where installation is a requirement. For RFPs requiring resumes, it is important to consider them as a vital factor within the proposal. It is absolutely necessary to tailor resumes to the prospective customer and to the RFP personnel and technical requirements. Many personnel may have a broad range of skills, capabilities, and experience. Remember however, that the Government client is interested primarily in the employee's skills and background that relate to the RFP requirements. Therefore, it is necessary that an employee's resume be truthfully written in such a manner as to assure the government that the proposed personnel possess the highest skills for performance in the job categories as specified in the RFP.

Typically, the Government will specify the format to be used for resumes. Experience has shown that the typical page count limits placed by the Government on technical proposals do not include the resume sections. There are exceptions of course, and the writers should always be aware of limitations. A major step to be taken in preparing resumes to meet the requirements of the various positions cited in the RFP labor categories is to utilize the list of key words wherever reasonable. This listing of key words or phrases would have been prepared during the requirements definition phase and must have been made available to all writers. A key role of the resume writer is to use the same words in the resumes that appear in the appropriate labor category descriptions provided by the Government. Remember that the Government's review and evaluation staff will be checking to determine whether or not the proposed personnel qualify in the labor categories listed, so make it easy for them to confirm that they do

If no resume format is specified within the RFP, the bidder will basically have a choice of formatting. It is essential that the employees resume contain as much information concerning his or her skills and background. Again, concentrate on skills that are required by the client as specified in the RFP. Section 16.1 shows a resume chronologically formatted that meets these needs and is recognized by everyone. The chronological resume is generally specified by the government in most RFPs.

16.1   Personalize the Resumes

Resumes present the background and capabilities of individuals within the firm. The term individuals cannot be over-emphasized. These are technically capable individuals, not manufactured products. Thus the human aspect is important. Individuals have names, faces, personalities, and skills important to the prospective client. They are the most important commodity in service-oriented acquisitions.

In personalizing the resumes, consider whether the employee has ever visited the prospective client's facility on a marketing venture. If so, this should be shown. Don't be afraid to state within the resume that the individual has stated an interest in supporting any contract that may be awarded as a result of the proposal. Clearly identify special skills the individual has such as leadership, client interface skills, and other important assets. Finally, it may aid the effort by providing a photograph of the individual within the resume. This adds reality to the proposal.

John Hansen, MS EE
University of Wyoming, 1995

Photogaphs add
faces to names

Mister Hansen (John) is an accomplished **Project Manager** in areas closely related to the requirements set forth by the Air Force in RFP 162. His professional background consists of 15 years of experience in Government contracting, specializing in **Project Management on multiple Government contracts.** He is expert in **software quality assurance** and other key requirements including client interface, teaching of software development and quality classes that he himself developed. John visited the Patrick AFB site earlier in the year and conducted technical discussions with the Air Force staff along with other members of FSI.

EXPERIENCE

2007 to Present   Federal Solutions Inc. – Engineering Manager

Mr. Hansen has recently completed a project for the US Air Force where he managed the development of a computer based **Independent Verification and Validation (IV&V)** program for intensive testing of the functionality of an Aircraft Maintenance Software System (AMSS) for use in C-130 transport planes. This system is used to verify that computer software, documentation, and maintenance records accurately mirror the maintenance processes that are performed on the aircraft. The **IV&V** system has been made available, and is being used by other agencies using C-130 aircraft. The experience he has gained in this program is directly related to the subject RFP.

On prior assignments, John has provided engineering support for a number of programs managed by FSI. These include spacecraft data processing design for many of NASA's unmanned spacecraft. Foremost of these is the Landsat spacecraft that provides images of the earth's terrains in a manner that environmentalists can monitor the earth's surface. He also designed and managed the development of a spacecraft weather monitoring system used by the National Weather Service (NWS) and multiple environmental monitoring agencies.

1999 to 2007   General Electric Space Division – Manager of Ground Support Data Systems

Here, John performed a wide range of Program Management and Engineering Management functions in support of NASA, DOD, and the Department of Agriculture. He managed the quality aspects of information data evaluation efforts relating to military imaging spacecraft and environmental spacecraft. His efforts produced final quality assurance plans used to test and evaluate spacecraft imagery from multiple satellites. He managed teams of quality analysts to evaluate and detect erroneous conditions within the processing of imagery data.

17. REVIEWS

Quality and Responsiveness are imperative!
Every proposal effort should include a
Series of reviews during the proposal development.

The number and type of proposal reviews vary among companies. To understand the overall review process, we will consider the minimum 5 that are necessary to cover the important bases. It is vital that the review process times be built into your development schedule.

Table 5 – In-Process Reviews

Name of Review	Process
Yellow Team Review	Confirms that the outline is consistent with the RFP Section L requirements. This team also reviews key words and phrases from the RFP that will be used by the writers.
Blue Team Review	Reviews expanded outline which now contains (quasi-hidden) requirements from other RFP sections such as J & M. Key words and phrases are checked again.
Pink Team Review/s	Reviews multiple (if necessary) submittals of the draft proposal. Final approval serves as input for the Red Team.
Red Team Review	Reviews final version as deemed passable by the Pink Team. This is the single most critical review.
Gold Team Review	Conducts final corporate review and approval – Includes Technical and Price

17.1 The Yellow Team

The initial document review is performed by the Yellow Team. This team performs two major functions. First is the review of the proposal outline as prepared by the writing team and the comparison of the outline with the instructions of RFP section L. The Government may have provided an outline for the proposal. If so, the Government's outline will prevail – at least in terms of major sections and content. The yellow team will also look for proposal key words in the outline titles. Errors detected are verified and corrected by the writers.

17.2 The Blue Team

Following the Yellow Team review, the outline will be corrected if necessary and expanded to include additional RFP sections that should be addressed in the proposal. Yes, it doesn't necessarily end with the Statement of Work from Section C. The team must be aware that requirements from additional sections of the RFP must be embedded in the outline. This will include RFP Sections J & M and RFP attachments. The Blue team will conduct its review and findings in the same manner as that of the Yellow Team.

17.3 The Pink Team

Next is the Pink Team review. It is important to note that many contractors use a multiple of review teams during the proposal development process. Companies simply vary in their procedures. For this document, five teams are described.

The Pink Team functions in a similar manner. In both cases, the reviews are intended to answer the following questions, in order of their priority to the Pink Red Team reviewers:

> ➤ Does the proposal satisfy all of the RFP specifications?

> ➤ Is the material technically correct?

The primary concern of the Pink Team is the responsiveness of the draft proposal to the technical and programmatic issues set forth in the RFP. That is, if it was in the RFP, was it addressed in the proposal? If the government evaluators can't find the material they're looking for in order to score you on each subject to be addressed, it's almost a certainty that they won't bother to wade through the technical material. If the draft is not responsive, the writing team may have a serious problem to correct depending on the degree of the inconsistencies. Because of the prior reviews by the Yellow and Blue teams working in concert with the writers, this should not happen. Even so, the Pink Team must assure that the material is responsive.

The second element of concern is for the correct technical responsiveness of the proposal. Specifically, the review team must ask: "Does the proposal respond correctly to the RFP"?

The Pink Team is required to become part of the solution, not part of the problem. As such, listen to what they have to say and accept their criticisms as well intended contributions. Having identified deficiencies, the burden is on them to suggest how the material can be improved. The Pink Team review/repair process will continue through multiple submissions until satisfaction in the written response is achieved.

17.4   The Red Team

The Red Team is the final review team that simulates, as closely as possible, the Government review process. This is the most critical review team of all because it evaluates the finished product. The Red Team will review the written proposal objectively, and "score" the proposal against the requirements of the RFP. Note that the previous three review teams have completed their job and the proposal has been deemed responsive. The Red Team attempts to simulate the customer's evaluation process and scores the proposal draft using review processes that are as thorough and as close as possible to those of the government client. The Team will:

> ➤ Review and evaluate the proposal from the government client's perspective. Is this contractor proposing to provide what we have asked for in the RFP?

> ➤ Identify errors and presentation weaknesses within the document. Strengths must also be identified.

> ➤ Provide written recommendations that will help the writing team correct and/or revise the document to make it more sellable.

> ➤ Assist the writing team to make corrections and/or clarifications needed. This does not involve actual writing, simply assistance and clarification. The Proposal Manager and Capture Manager decide which Red Team criticisms are valid and will thus be used as proposal corrections.

The Red Team leader is identified early in the proposal process. He or she will help to identify other members of the team and will perform the review along with other members of the team. The Red Team leader guides the review team by providing instructions and coordination during the review process. He or she shall:

- ➢ Consolidate comments and recommendations
- ➢ Adjudicate differences.
- ➢ Perform the briefing of Red Team findings to the proposal writing team.
- ➢ Assist in providing rationale for the acceptance of any Red Team comments by the key responsible managers.

The evaluation procedure requires each Red Team member to score all assigned sections or chapters. There are a wide range of "faults" that the Red Team may identify during its review. It is also true that different individuals will detect different faults. The following list contains some of the faults that may be individually identified.

- ➢ Unsupported statements and conclusions.
- ➢ Doesn't state advantages relative to competition.
- ➢ Lacks embedded themes.
- ➢ Lacks an innovative approach.
- ➢ Inadequate discussion of concentration on high-risk areas found in the RFP.
- ➢ Improperly assumes customer has prior knowledge of the company, its capabilities, and its experience, thus, fails to display the full resources of the company.
- ➢ Gold-plating: that is, it contains features beyond minimum requirements at no additional cost.
- ➢ Missing or buried conclusions.

A SCORING MECHANISM

The Red Team review is of course crucial. I strongly suggest a scoring method applied by each member of the review team. Realizing that few people actually think alike, it is important that an evaluation standard be implemented. The standard will not make everyone think alike, but it will get the review team closer to uniform evaluation and grading. It is important to note that while any imposed standard will not solve all of these thinking standards, it will get us closer. I suggest that we apply an evaluation standard that is simple. The Red Team leader should impose the standard that is based on previously listed items for the team to review. A simple scoring mechanism will be imposed on the writing by each member of the review team. This simple process shown in Figure 12 is productive.

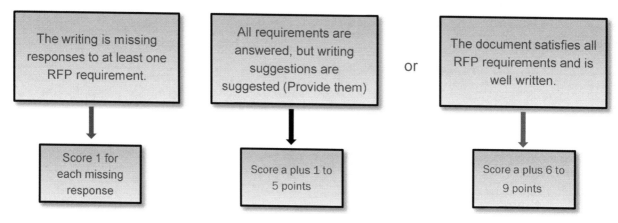

Figure 12 – A Scoring Process

Your scoring will be two numeric characters. The range of these scores is 0 through 9. If the score begins with a number other than zero (0) there is a problem/s that must be repaired. It is unlikely that this will occur because the Pink Team should have identified the error during its review. Scores between 01 and 09 should be acted on as the score number implies.

PROPOSAL TEAM DEBRIEFING

A portion of the last day of the review will be spent by the Red Team in debriefing key members of the proposal team. At a minimum, the debriefing will cover:

➤ General comments

➤ Major strengths and weaknesses by chapter

➤ Identification of focus requirements (clarity, presentation, etc.) by chapter

➤ Positive fixes and recommendations for improvements by chapter

➤ Suggestions for material to be included in the Executive Summary.

The suggested format for the debriefing is for the Red Team leader to present an overview of the Red Team findings. Individual Red Team members should be prepared to present and discuss his or her review comments more specifically. But, these discussions need to maintain focus.

It will be the responsibility of the Proposal Manager to determine if and how the various Red Team comments are incorporated into the final proposal. The proposal manager will document the manner in which this was done.

Finally, the Proposal Manager has the final word on accepting or rejecting any or all Red Team comments.

17.5 The Gold Team Review

The Gold Team consists of the company's executives. This review is critical in that it commits the corporation to all technical, management, and legal terms of conditions of the proposal.

18. THE COST PROPOSAL

We have all heard it said, "It all comes down to price". In the world of government proposal creation, that phrase is only true when we are bidding on Firm fixed-Price contracts. All other types of government contracts allow the selected contractor to retrieve contract performance costs plus varying fees (profits). That said, contractors should always price to win. In Section M of the RFP, the reader may see the percentage of the overall evaluation that is assigned to price, technical, and management sections. In the paragraphs to follow, multiple contract types are discussed. But first, let's look at the general cost proposal structure.

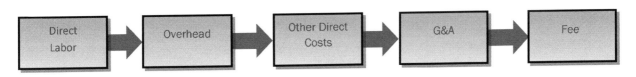

Figure 13 – Cost Proposal

18.1 Element Structure

18.1.1 Direct Labor

The salaries you pay to proposed employees

18.1.2 Overhead

The cost of employee benefits, office space, furniture, office equipment and other corporate costs

18.1.3 Other Direct Costs

The Material and Products purchased by the government. This also includes consultants who are proposed to augment direct labor employees.

18.1.4 General & Administrative

Corporate Officer salaries, office personnel salaries, legal and other corporate costs of doing business. These costs are pro-rated across contracts.

18.1.5 Fee

The profit to the corporation

Additional and highly amplified information on the costing structure for Government contracting may be found in Federal Acquisition Regulation (FAR) Subpart 31-2, Contracts with commercial organizations. Other FAR elements will also apply.

It is highly recommended that cost proposal support for the winning of government contracts be acquired by your corporation. These experienced individuals are necessary in any Government contractor's organization to establish financial operations including all contracting matters. There is no success without legal and contract support in your organization, be they employees or on an as needed basis.

18.2   Types of Government Contracts

18.2.1   Fixed Price Contracts

You must price your fixed price proposal not only to win, but to earn profit!

The term Firm Fixed Price is simply what it says. The bidder evaluates the RFP and calculates the cost to the corporation, inclusive of all elements of Labor, Overhead, Other Direct Costs G&A and Fee. The winning firm must provide the proposed materials and labor for the price proposed. There are no ifs, ands, or buts about it. No matter what the actual cost is to the firm, the Government will pay no more than the award price. This approach is typically used by the government in the RFP only where the purchase can be clearly and completely identified. There are three major types of Firm Fixed Price Contracts:

➢ *Firm Fixed Price - FFP* – Firm Fixed Price As noted above, Firm Fixed Price contracts contain highly detailed requirements. This type of contract requires that the firm closely manage the costs if the contract is to be profitable. Higher than proposed costs equals less fee or possibly even a loss. Lower costs equal a profit to be realized.

➢ *Fixed Price Contract with Incentive - FPIF* – The fixed price contract with incentive fee is a firm fixed price type contract, except that the proposed fee segment may vary. The variation depends on the contractor's cost performance. If the contractor keeps prices below the bid level, additional fee will be realized.

➢ *Fixed Price with Economic Price Adjustment* – These are fixed price contracts that contain a provision to adjust the contract fee predicated on cost changes outside of the contractor's control.

Fixed price contracts may include labor only, materials only, or labor and materials. The Government may specify the number and detailed description of labor and/or products to be purchased. However, many RFPs may provide detailed functional requirements and leave the number of personnel and equipment needed to satisfy contract requirements to the specific bid of the contractor. In this manner, contractors are required to prepare their specific solution to the government in the contractor's solution to the requirements.

A sample typical proposal may include say, the number of personnel plus materials to keep a government office building clean. As can be seen, the government is providing the requirement. The bidding contractors are providing their approach based on their site visits to the government's facility. While this may appear to be simple, each contractor in its proposal will specify its approach to the solution. Based on this example, the contractors may vary in the number of personnel and materials they require to satisfy the requirement. Keep in mind that the contractor must satisfy the contract needs with the number of personnel and materials bid, or face a loss of dollars.

The fixed price contract cited here while simple in the contractor's approach, includes the need for safeguards by the contractor. For instance, consider the case of a multi-year contract. In its proposal, the contractor may be required to provide pricing for five years of contract duration. Salary increases must be considered for employees as well as the cost increases of cleaning products during later years. The resulting contract award for this proposal will nearly always be based on the total bid price for the five year period.

18.2.2   Cost-Plus Contracts

Cost Plus contracts may vary in their execution. These types of contracts carry far less risk to the contractor than fixed price contracts. For our purposes consider basically three (3) types of Cost Plus contracts that the Government may award. These are:

Cost Plus Fixed Fee - CPFF:

A CPFF contract reimburses the contractor for the cost incurred to complete the work plus fixed fee negotiated at the time of award. In this type of contract, the word Fixed is key. The fee is negotiated at award time based on the estimated costs. This fee will remain the same no matter what the costs turn out to be. Thus, based on a contract cost overrun, the contractor will make no fee based on the cost overrun. None the less, the CPFF contract is highly favored by many contractors because the financial risk is low.

Cost Plus Incentive Fee – CPIF:

Under the Cost Plus Incentive approach, the contractor is paid a higher fee if the cost can be kept below the proposed estimate. This is calculated by a formula identified by the government. This approach encourages the contractor to keep costs down.

Cost Plus Award Fee CPAF:

A cost reimbursement contract where the objectives of the contract are determined to be completed by subjective means. This occurs many times because the contract requires elements of research, such as in very high technology work. The contractor receives reimbursement for their costs plus the award fee. The resultant fee is determined by the contracting office based on results achieved by the contractor. Cost plus award fee contracts are not normally used by the Government when a cost plus fixed fee or cost plus incentive fee contract would be more appropriate.

Cost plus contracts are common within the Government. These types of contracts provide flexibility as needs change. For instance, the Government may require additional equipment based on increases in workflow or when establishing an additional facility. These types of changes may also require additional personnel to be supplied by the contractor. This of course requires a contract change which is commonplace in these types of contracts.

18.2.3   Labor Hour Contracts

Labor Hour contracts are highly desirable from both a government and contractors perspective. These contracts, as well as Cost Plus contracts, offer the government the capability to easily modify the contract through the addition of personnel when overall needs change. Typically, the number of base award personnel does not change in a downward manner during a specified time period. Increases in personnel may be requested at any time by the government as need arises. This type of contract modification increases all elements of the price structure.

18.3   The Cost Proposal

Remember that a great deal of thought and planning must be performed in the process of developing your price. While this section is of limited use to those who have been called upon to write the technical and management sections of the proposal, there is a need by others in the company to understand the basics of this information. Section M of the RFP, typically contains the percentage of the overall evaluation that is applied to price, technical, and management. Incidentally, this information is normally isolated from the proposal writing team. It is typically completed by the legal and financial officers and the corporate management team.

We have all heard it said, "It all comes down to price". In our world of government proposal creation, that phase is particularly true when bidding on firm fixed-price contracts. All other types of government proposal pricings allow the winner to retrieve contract costs plus varying fees (profit). You should always price to win. Because of risks, many government contractors do not bid on firm fixed price contracts. Consider the following.

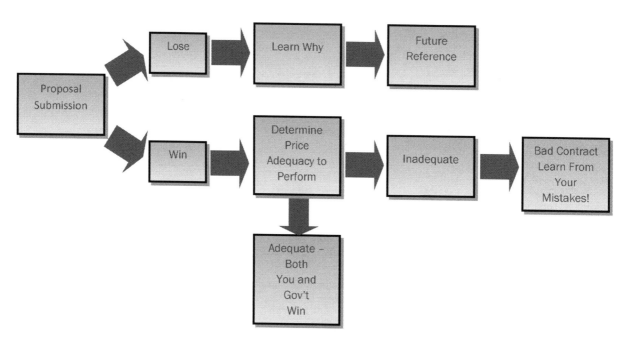

Figure 14 – Ensure Pricing is adequate on Fixed Price Contracts

19.   CONTRACT SET-ASIDES

19.1   Small Business Act

Founded in 1953 by congress through passage of the Small Business Act, the Small Business Administration (SBA) has grown to its current role of ensuring that small businesses share in the contract spending of our government. Theirs was not an easy role. Consider the number of major corporations that existed back in 1953. As far as government buyers were concerned, it made perfect sense to buy from these large businesses. How could the government go wrong in this approach? After all, they thought; Should the Government not buy from the best known companies?

Today, it is hard to imagine government contracting on simply a full and open competition of all vendors. Small businesses would not stand a chance against these major businesses that offer everything imaginable. Our Government tackled this problem through the continued growth of the SBA. Over the years, they have continuously modified programs to help ensure that all small businesses in America had the opportunity to succeed. The approach used by the SBA is to create set-asides of government contract spending to target contracts to qualified small businesses of many types. The government has also mandated that large businesses set targets of contract spending in the direction of small businesses on a sub-contracting basis. This area in the belief of this writer, needs more scrutiny.

19.1.1   The 8(a) Business Development Program

In looking at the contract set-aside system, we will see that the government has set-aside opportunities for many types of small businesses. Foremost among these is the 8(a) program used to set contracting opportunities aside for firms that are determined to endure racial discrimination in their efforts to win contracts.

The 8(a) Business Development Program is a business assistance program for small disadvantaged businesses. The 8(a) Program offers a broad scope of assistance to firms that are owned and controlled at least 51% by socially and economically disadvantaged individuals.

Participation in the program is divided into two phases over nine years: a four-year developmental stage and a five-year transition stage. Access to approval into the 8(a) program can be long and frustrating on behalf of the applicant firm.

Benefits of the Program

> ➢ Participants can receive sole-source contracts, up to a ceiling of $4 million for goods and services and $6.5 million for manufacturing.

> ➢ 8(a) firms are also able to form joint ventures and teams to bid on contracts. This enhances the ability of 8(a) firms to perform larger prime contracts and overcome the effects of contract bundling, the combining of two or more contracts together into one larger contract.

In addition, 8(a) participants may take advantage of specialized business training, counseling, marketing assistance, and high-level executive development provided by the SBA and its resource partners. 8(a) firms may also be eligible for assistance in obtaining access to surplus government property and supplies, SBA-guaranteed loans, and bonding assistance for being involved in the program.

For general questions about the 8(a) Business Development program, the reader may contact 8aquestions@sba.gov (link sends e-mail). This address is valid at the time of this writing.

19.1.2   Small Disadvantaged Business

Small businesses can self-represent their status as a Small Disadvantaged Business (SDB). SDB firms are separate from 8(a) firms and receive less of the business from government coffers.

You do not have to submit an application to SBA for SDB status.

To self-represent as an SDB, register your business in the System for Award Management. However, you and your firm must still understand the SBA eligibility criteria for SDBs. Generally, this means that:

> The firm must be 51% or more owned and control by one or more disadvantaged persons.

> The disadvantaged person or persons must be socially disadvantaged and economically disadvantaged.

> The firm must be small, according to SBA's size standards

The reader should note that in addition to self-representing its business as an SDB, if qualified, the firm may also meet the requirements for any of the following programs:

> SBA's Small Business Development Program may provide managerial, technical, and contractual assistance to small disadvantaged businesses to ready the firm and its owners for success in the private industry.

> SBA's HUBZone Program helps small businesses in urban and rural communities gain preferential access to federal procurement opportunities. A geographical area is designated as a hub zone by the government because of the lack of businesses in this financially depressed area. These preferences go to small businesses that obtain HUBZone certification in part by employing staff who live in a HUBZone. The company must also maintain a "principal office" in one of these specially designated areas. For further description of HUBzone procurements, see section 1.9.1.5.

19.1.3   Women-Owned Small Business

The Women-Owned Small Business Federal Contract Program authorizes contracting officers to set aside certain federal contracts for eligible women-owned small businesses. In 2011, the SBA published a final rule aimed at expanding federal contracting opportunities for women-owned small businesses (WOSBs). The Women-Owned Small Business (WOSB) Federal Contract program authorizes contracting officers to **set aside certain federal contracts** for eligible:

> Women-owned small businesses (WOSBs) or

> Economically disadvantaged women-owned small businesses (EDWOSBs)

To be eligible, a firm must be at least 51% owned and controlled by one or more women, and primarily managed by one or more women. **The women must be U.S. citizens**. The firm must be "small" in its primary industry in accordance with SBA's size standards for that industry. In order for a WOSB to be deemed "economically disadvantaged," its owners must demonstrate economic disadvantage in accordance with the requirements set forth in the SBA's standards.

WOSB Program Third Party Certification

Women Owned Small Businesses may elect to use the services of a Third Party Certifier to demonstrate eligibility for the program, or they may self-certify using the SBA's approved process. SBA only accepts third party certification from these entities, and firms are still subject to the same eligibility requirements to participate in the program.

There are four third Party Certifiers under the WOSB Program. The four organizations are:

> **El Paso Hispanic Chamber of Commerce**

  2401 E. Missouri St. El Paso, TX 79903

> **National Women Business Owners Corporation**

  1001 W. Jasmine Drive, #G Lake Park, Florida 33403

> **US Women's Chamber of Commerce**

  700 12th St. NW, Suite 700, Washington D.C.

> **Women's Business Enterprise National Council (WBENC)**

  1120 Connecticut Avenue, NW, Suite 1000 Washington, DC 20036

19.1.4   Service-Disabled Veteran-Owned Businesses

The Service-Disabled Veteran-Owned Small Business Concern Procurement Program provides procuring agencies with the authority to set acquisitions aside for exclusive competition among service-disabled veteran-owned small business concerns.

The purpose of this program is to provide procuring agencies with the authority to set acquisitions aside for exclusive competition among service-disabled veteran-owned small business concerns, as well as the authority to make sole source awards to service-disabled veteran-owned small business concerns.

Eligibility

To be eligible for the SDVOSBC, you and your business must meet the following criteria:

> The Service Disabled Veteran (SDV) must have a service-connected disability that has been determined by the Department of Veterans Affairs or Department of Defense
> The SDVOSBC must be small under the North American Industry Classification System (NAICS) code assigned to the procurement
> The SDV must unconditionally own 51% of the SDVOSBC
> The SDVO must control the management and daily operations of the SDVOSBC

19.1.5   HUBZone Procurements

The SBA's **Historically Underutilized Business Zones** (HUBZone) program was placed into service by the U.S. Small Business Administration. The program encourages economic development in historically underutilized business zones. These zones are typically poor neighborhoods. This means that your business must be headquartered in a historically underutilized community as determined by the SBA. This program and geographically qualified areas (zones) are determined by the SBA who regulates and implements the HUBZone program.

The program's benefits for HUBZone-certified companies include:

➤ Competitive and sole source contracting

➤ 10% price evaluation preference in full and open contract competitions, as well as subcontracting opportunities.

The federal government has established a goal of awarding 3% of all dollars for federal prime contracts to HUBZone-certified small business concerns.

To qualify for the program, a business must meet the following criteria:

➤ It must be a small business by SBA standards

➤ It must be owned and controlled at least 51% by U.S. citizens, or a Community Development Corporation, an agricultural cooperative, or an Indian tribe

➤ Its principal office must be located within a "Historically Underutilized Business Zone".

➤ At least 35% of its employees must reside in a HUBZone.

19.2   Determining Business Size

One of the first steps in becoming a government contractor is to accurately determine if you can qualify as "small" under the SBA size standards. In other words, you must be defined as a small business when submitting proposals for small business contracts.

SBA uses the North American Industry Classification System (NAICS) as the basis for its size standards. To determine your firm's qualification as a small business, you may visit the official NAICS website to find the code(s) that apply to your industry. You may access this site at www.census.gov/eos/www/naics.

19.3   Mandatory Registrations

Prior to bidding on Government contracts, contractors must be registered in several organizations. These are one-time registrations which may from time to time be updated because of changes in your firm.

19.3.1   DUNS Number

You must obtain a DUNS number, a unique nine-digit identification of your firm. You will need a separate DUNS number for each physical location of your business.

DUNS numbers are acquired at no cost to the contractor through Dun & Bradstreet. Again, this number/s is mandatory for all businesses who wish to submit proposals to win federal government contracts.

When registering for your D-U-N-S Number, you will need the following on hand:

> Legal name

> Headquarters name and address for your business

> Doing Business As (DBA) or other name by which your business is commonly recognized

> Physical address, city, state and ZIP Code

> Mailing address (if separate from headquarters and/or physical address)

> Telephone number

> Contact name and title

> Number of employees at your physical location

> Whether you are a Home-Based Business

You may acquire your D-U-N-S Number by visiting the Dun and Bradstreet web site at:

http://www.dnb.com/get-a-duns-number.html

19.3.2   The System for Award Management

The System for Award Management (SAM) is the Official U.S. Government system that consolidated the capabilities of CCR/FedReg, ORCA, and EPLS. There is NO fee to register for this site. Entities may register at no cost directly at **http://www.SAM.com**   User guides and webinars are available under the Help tab.

Your firm cannot do business with the Federal Government without this registration!

19.3.3   Tax I.D.

You must have a Tax ID in order to conduct business with the Federal Government. An Employer Identification Number (EIN) is also known as a Federal Tax Identification Number, and is used to identify a business entity. Generally, businesses need an EIN. You may apply for an EIN on line by visiting www.gov-tax.com .You should also check with the state in which you are headquartered to determine if you need a state number or charter.

19.3.4   Cage Code

The **Commercial and Government Entity** Code, or **CAGE Code**, is a unique five (5) digit identifier assigned to suppliers to various government or defense agencies. CAGE codes provide a standardized method of identifying a given facility at a specific location.

CAGE Codes are used internationally as part of the NATO Codification System (NCS), where they are sometimes called **NCAGE Codes**. CAGE codes are referenced in various databases of the NCS, where they are used along with the supplier's part number to form a reference which is held within the National Stock Number (NSN) record. This reference enables users of the NCS to determine who supplies any given part.

To acquire a CAGE Code,

> ➢ Visit: https://www.sam.gov

> ➢ Register in SAM

> ➢ CAGE is assigned to your company as part of SAM validation process

> ➢ Once registration is active, log into SAM account

> ➢ View your CAGE code

19.3.5 Contractor Performance Assessment Reporting System (CPARS)

CPARS registers contractor performance. This system hosts a suite of web-enabled applications that are used to document contractor and grantee performance information that is required by Federal Regulations.

FAR Part 42 identifies requirements for documenting contractor performance assessments and evaluations for systems, non-systems, architect-engineer, and construction acquisitions. FAR Part 42 also requires documenting additional contractor performance information in the Federal Awardee Performance & Integrity Information System (FAPIIS), including Terminations for Cause or Default, DoD Determination of Contractor Fault and Defective Cost or Pricing Data and to make the information available in the Past Performance Information Retrieval System (PPIRS).

FAR Part 9 identifies requirements for Contracting Officers to enter Determinations of Non-Responsibility in FAPIIS. The Grant Community is also required to utilize FAPIIS to document Terminations for Material Failure to Comply and Recipient Not Qualified Determinations.

The CPARS applications are designed for UNCLASSIFIED use only. Classified information is not to be entered into these applications.

To visit CPARS, go to: https://www.cpars.gov/index.htm

19.4 Helpful Government Contracting Sites

There are a number of government related agencies that are helpful to contractors. While not necessarily mandatory, contractors will find the information provided or offered, to be very helpful in the quest to locate and acquire federal contracts. Find listed below a number of helpful websites that contain all of the instructions and guidelines necessary to becoming a successful government contractor.

> ➢ Access the FAR http://acquisition.gov/far/index.html?menu_id=40
> This is the internet home page for federal acquisitions. All you need to know concerning Government regulations for purchasing is here. This is typically used only as a search for details of specific subjects in the regulations.

> ➢ Code of Federal Regulation (13CFR). The web addresses found here point the reader to an extensive list of information of interest to small businesses.

> ➢ Code of Regulations Chapter 1, Part 121 presents the standards and regulations of the Small Business Administration. http://www.ecfr.gov/cgi-bin/text-idx?tpl=/ecfrbrowse/Title13/13cfr121_main_02.tpl . This document will be of interest to minority contractors for qualifications purposes.

> ➢ Federal Business Opportunities http://www.fbo.gov is a daily listing of all Federal Procurements in various stages of acquisition. It is an excellent site to view upcoming opportunities or to view

existing contracts and the associated contractors.

➢ GSA Schedules http://www.gsa.gov/portal/content/197989. This website is provided to allow any contractor to qualify as an approved source to win competitive General Services Administration contracts. This is in the process of changing to allow multiple registrations on a single site. I suggest readers visit this website to learn more.

➢ GWACS http://www.gsa.gov/portal/content/104874. This is an on-line training document presented in Adobe Acrobat format. This particular site presents training on how to sell materials and services. There are two additional classes available on separate web sites.

➢ http://www.gpo.gov/fdsys/browse/collectionCfr.action?collectionCode=CFR. This is a training manual on Government contracting.

➢ Local PTAC http://www.dla.mil/SmallBusiness/Pages/ptac.aspx. PTAC is an acronym for procurement technical assistance centers. These exist at Air Force facilities and are designed to assist small businesses in acquiring Government business. Large companies are also in contact with PTAC to learn of small businesses and their potential teaming.

➢ North American Industry Classification System http://www.census.gov/eos/www/naics/ The U.S. Small Business Administration (SBA) is amending its Small Business Size Regulations to incorporate the Office of Management and Budget's (OMB) 2012 modifications of the North American Industry Classification System (NAICS), identified as NAICS 2012, into its table of small business size standards. This material is important to small businesses to maintain their size qualifications.

➢ Office of Women's Business Ownership http://www.sba.gov/about-offices-content/1/2895 At this website, the Small Business Administration presents listings of additional websites that may well be of interest to Woman owned businesses.

➢ Procurement Technical Assistance Centers http://www.aptac-us.org/new/Govt_Contracting/find.php. This is a commercial website that assists small business by training them for success.

➢ SAM www.sam.gov. The **System for Award Management** (SAM) is the Official U.S. Government system that consolidated the capabilities of CCR/FedReg, ORCA, and EPLS. There is NO fee to register for this site.

➢ SBA Government Contracting http://www.sba.gov/about-offices-content/1/2986. This important website walks the reader through the contracting elements the business will need to become a federal contractor. It identifies elements such as Standard Industrial Classification (SIC) codes, federal tax I.D. acquisition, and other elements needed to become a federal contractor. It is an excellent site for small businesses to visit so that the elements required to succeed may be understood,

➢ BA Size Standards http://www.sba.gov/content/am-i-small-business-concern . The question pursued and answered in this website is whether or not your firm is in fact a small business firm by government standards. The SBA, for most industries, defines a "small business" either in terms of the average number of employees over the past 12 months, or average annual receipts over the past three years. In addition, SBA defines a U.S. small business as a concern that:

➢ Is organized for profit

➢ Has a place of business in the US

➢ Operates primarily within the U.S. or makes a significant contribution to the U.S. economy through payment of taxes or use of American products, materials or labor

- Is independently owned and operated
- Is not dominant in its field on a national basis

This in fact is interesting reading and like many other government documents listed here, it helps you with your decision making along the way to becoming a federal contractor.

- SCORE http://www.score.org/chapters-map, The SCORE Association "Counselors to America's Small Business" is a nonprofit association comprised of 13,000+ volunteer business counselors throughout the U.S. and its territories. Our Government trains SCORE members to serve as counselors advisors and mentors to aspiring entrepreneurs and business owners. These services are offered at no fee, as a community service.

- Small Business Development Centers http://www.asbdc-us.org.

  America's SBDC represents America's nationwide network of Small Business Development Centers (SBDCs). The mission of America's nationwide network of SBDCs is to help new entrepreneurs realize the dream of business ownership, and to assist existing businesses to remain competitive in the complex marketplace of an ever-changing global economy.

  Hosted by leading universities, colleges and state economic development agencies, and funded in part by the United States Congress through a partnership with the U.S. Small Business Administration, nearly 1,000 service centers are available to provide no-cost business consulting and low-cost training.

- Standard Forms SF 33, SF 1449, SF 1447, SF 18, SF 26 http://www.gsa.gov/portal/forms/type/SF This Government website provides directions to acquire various forms that are used in proposals in response to Government RFPs.

- SUB-Net http://web.sba.gov/subnet/search/index.cfm . There are a number of internet websites from locations in many states of the country. These sites provide the reader with information on subcontracting opportunities.

- Women's Business Centers http://www.sba.gov/content/womens-business-centers. The Women's Business Centers (WBCs) represent a national network of nearly 100 educational centers throughout the United States and its territories, which are designed to assist women in starting and growing small businesses. WBCs seek to "level the playing field" for women entrepreneurs, who still face unique obstacles in the business world. SBA's Office of Women's Business Ownership (OWBO) oversees the WBC network, which provides entrepreneurs (especially women who are economically or socially disadvantaged) comprehensive training and counseling on a variety of topics in several languages.

Yes, the list is long. But remember that they exist to help contractors become successful. Some of these are partially repetitive in some aspects but make no mistake, they are helpful.

20. THE COVER LETTER

The purpose of the cover letter is to voice the Corporation's commitment from the highest level.

The cover letter for the technical and management volumes is sometimes called the "President's Letter." Constrained to few pages of letterhead paper, the letter usually consists of material on the following items:

> ➤ A complete reference to the solicitation, including title, solicitation number, and date.

> ➤ A reference to the importance of the program addressed and to our proposal being totally responsive to the solicitation. (Anything less than a proposal that is totally responsive to the solicitation has a good chance of not being evaluated.)

There may be a valuable reference to actions taken by the letter writer for the benefit of the government. Such actions might include one or more of the following:

> ➤ Appointment of a key person as FSI's Program Manager (presumably, the customer has already met this person and would benefit from this individual on the team!)

> ➤ A directive that only our most experienced personnel will be assigned to the resultant contract.

> ➤ Assurance that the FSI Inc. Program Manager will have priority access to Corporate resources.

The letter should then conclude with statements to the effect that the firm is fully committed to support the government, that the writer is available to talk with the Government at any time regarding the proposed effort (include the writer's office telephone number), and that FSI is ready to begin work immediately upon award of contract.

21. COVER ART BINDERS

Even before one word is read, the evaluator will draw a conclusion about your proposal based on the cover art and type of binding used. This being the case do not scrimp on either.

With respect to the cover art, choose or develop a piece of artwork that goes to the heart of the problem ("the theme") and that will be of interest to the evaluators. The corporation's or team's name or logo should be shown prominently in either the upper (preferred) or lower portion of the cover, with the formal name of the solicitation title and name also displayed. Use the same cover for the technical, management, and cost volumes, being careful to label them appropriately.

With respect to bindings, most large volumes are bound in three-ring binders with insert covers for the front, back, and spine. Experience has shown that so called "D-ring binders are preferred by Government evaluators because they lie flat when opened. The availability of pockets on the inside of the front and back covers may also be necessary if there is a need to include various items with the various volumes (e.g., a multi-page Executive Summary).

Federal Solutions Incorporated

QUALITY CONTROL AND SYSTEM QUALIFICATIONS DETERMINATION

TECHNICAL PROPOSAL

IN RESPONSE TO

SOLICITATION XYZ-BX-0984605

SUBMITTED BY

Federal Solutions Inc.

29 February, 2014

SUBMITTED TO

UNITED STATES AIR FORCE

AIR RESCUE OPERATIONS

CHATAHOOCHIE WISCONSIN

22. PROPRIETARY DATA

The Freedom of Information (FOI) Act permits nearly everyone to request information about our proposals from the Government customer. Whole proposals may be sent to a requestor by the sponsoring Government agency.

Freedom of Information requests can be (and, in some cases, are) frivolous. The requestor sees a title of a winning proposal in the Commerce Business Daily (CBD) and writes the agency for a copy just to read up on the technology. Some companies routinely request each winning proposal in technical areas where they are marketing. They learn who is in the field, what they are doing, and, in some cases, a lot of information about their competition.

As a matter of policy, firms should enact guidelines to cover claims of proprietary status for proposals. All three volumes of our proposals - Technical, Management, and Cost - should be marked as being proprietary.

A simple method to handle the notice on proprietary data is to preprint a notice at the bottom of each page of all volumes. The notice should read to the effect that:

"Use or disclosure of proposal data is subject to the restriction on the title page of this proposal."

The notice that is placed on the title may well read:

> **The data contained in all pages of this proposal shall not be used or disclosed except for evaluation purposes; provided that if a contract is awarded to this submitter as a result of or in connection with the submission of this proposal, the Government shall have the right to use or disclose this data to the extent provided in the contract. This restriction does not limit the Government's right to use or disclose any data obtained from another source without restriction.**

The above notwithstanding, some government agencies advise us when they receive a request for our proposal from a competitor. We may then be asked to tailor our proposal material so that some portions of it can be sent to the requestor. It is not at all clear just how much of our material we can have held back, despite whatever proprietary notice we have included in our proposal. The RFP Contracts Office should be contacted for their guidance regarding this subject. They have dealt with a number of agencies and have some experience to guide you.

23. ORAL TECHNICAL PRESENTATIONS

Be prepared for an oral technical presentation of your proposal.

Depending on the customer's specific requirements, you may or may not be requested to defend your approach. The odds are good that proposals involving honest competition, and representing a significant portion of the customer's budget, will require such a presentation.

Remember these important observations in getting ready for this presentation: The government's representatives have read and evaluated all of the proposals. They have selected a few (normally, not over three; more usually, two) for a final appraisal using oral briefings as a means by which to select the winner. Further, they have probably discussed the various proposals among themselves and their superiors. So, by this time, they have formed their own opinions.

Because of the general rules and guidelines on ethical conduct, it's almost a certainty that no information whatsoever on the government's deliberations will leak out;" nor should any be solicited. Instead, your firm's key personnel should put their heads together and decide:

> ➢ Why should the customer favor you over another firm?
>
> ➢ What are the strongest parts of our proposal?
>
> ➢ What are weakest parts of our proposal?
>
> ➢ Are we (favorably or unfavorably) unique?
>
> ➢ What would it take to convince them that we are the best?
>
> ➢ Who should we bring to the briefing as far as they are concerned?
>
> ➢ Are there any irritants in our proposal? Why?
>
> ➢ Who will be at the briefing for the customer? Do we know them?

Notice that there is no mention of money in the above list. If any information is volunteered about money, be alert and take notes. Generally, it is better to allow the customer to broach the subject. (If money is an issue, it will be brought up by the government's Contracting Officer.)

A briefing normally provides you with an opportunity to do some or all of the following:

> ➤ To clarify technical points in our proposal that the customer does not understand.

> ➤ To answer challenges to our technical solutions by a customer who does not believe our approach is the best and who needs to be convinced.

> ➤ To emphasize our good points (assuming we know what they are) and, we hope, to reduce the impact of our weak points. We can also reinforce our solution, our grasp of the problem, and our capabilities.

> ➤ To emphasize noteworthy points. If we can think of a real zinger to spring on the customer, now is the time. For example, "We have just been awarded a commendation by DoD for our performance on the XYZ program." Or: "Our proposed Program Manager has just received his MSEE degree."

The impression that you make at the oral briefing could be the deciding factor in the final award, assuming our price is not unfavorable.

24. POST AWARD DEBRIEFING

If we win a contract, who cares why we won? Well, you should!

If you know why we won, you can do it again, using your winning techniques on your next proposal. If you didn't win, then you will want to know why. In either case, a formal or informal debriefing is very useful to winning future work.

If the customer doesn't want to tell you why you lost, there is nothing you can do in a practical sense to force him. A debriefing held after a loss is useless unless the client really reveals the truth. Your actions during the oral presentations (if they are held) and, especially, your firm's perceived attitude and actions following a loss, color the customer's willingness to be frank and open at such a debriefing.

An informal debriefing is just that. It can be conducted by phone or on a short visit to the COTR's office. A formal debriefing should also be requested by your contracts administrator through the Government's contracting office. In every case, debriefings are only requested after an award has been made.

After a win, the best debriefing is an informal chat with both the COTR and the Government's Contracting Officer. Just ask what we did that was influential in our winning. This visit is usually a pleasant experience on the part of the customer, so he will normally be rather vocal. Be aware, however, that some surprises have surfaced in this type of discussion. For example, in a few cases (very few), companies learned that they won despite their proposal, meaning that in the evaluation criteria, your firm came out on top, even though your technical approach was not the best.

Copies of winning proposals can be ordered through the Government's contracts office by your contracts administrator. The Freedom of Information Act will allow us to receive these, although certain portions may be deleted by the customer as being proprietary to the winning firm. Reading these proposals often reveals a number of new techniques and ideas that can help us when we write our next proposal to the same or to a different customer.

25. IN CLOSING

Now, proposal writing to win Government contracts should no longer be a mystery to you. You are ready for the assignment as a writer. The key here is to blend with the writing staff as quickly as possible. Talk to them. You now have knowledge of what goes into typical Government proposals. The details of the processes used by your firm may vary somewhat from those that I have presented. None the less, the writing is the same. You should now have the confidence to perform. You know the subject matter, and you may now blend the matter with the process. Good writing everyone.

APPENDIX A

There are a wide number of federal contractor conferences that occur on a fairly frequent basis. Below, the reader will find just a few of these that may well be of importance. Contractors have a need to integrate their businesses in the overall government procurement industry. It is suggested that individual firms search the internet to find conferences of interest to their firm. This is an excellent approach to having your firm recognized by Government agencies. Below are listed a few of the many conference organizations that are important to government contractors.

1.   The Government Procurement Conference

The FBINC is a national conference fostering business partnerships between the Federal Government, its prime contractors, and small, minority, service-disabled veteran-owned, veteran-owned, HUBZone, and women-owned businesses. Now in its 25th year, the Government Procurement Conference has become the premier event for small businesses throughout the United States.

Participating firms will have the benefit of marketing their products/services to procurement representatives and small business specialists from government agencies. Companies may choose to set up an exhibit table to showcase their capabilities or simply come as an attendee.

The conference also includes educational conference sessions, procurement matchmaking, and a dynamic exhibitor showcase. Visit http://www.Fbinc.com

2.   GovEvents

The Government Procurement Conference is a national conference fostering business partnerships between the Federal Government, its prime contractors, and small, minority, service-disabled veteran-owned, veteran-owned, HUBZone, and women-owned businesses. Now in its 25th year, the Government Procurement Conference has become the premier event for small businesses throughout the United States.

Participating firms will have the benefit of marketing their products/services to procurement representatives and small business specialists from government agencies. Companies may choose to set up an exhibit table to showcase their capabilities or simply come as an attendee.

The conference also includes educational conference sessions, procurement matchmaking, and a dynamic exhibitor showcase. Visit http://www.Govents.com

Made in the USA
Middletown, DE
02 July 2020